Journal of Approximation Theory and Applied Mathematics

ISSN 2196-1581

Impressum:

Michael Rasguljajew

Kasinostraße 63
64293 Darmstadt
Germany, Hessen

E-Mail: m.rasguljajew@jatam.de

Website: www.jatam.com

Herstellung und Verlag:
BoD - Books on Demand, Norderstedt
ISBN 978-3-7357-9148-1

Contents

*Parameter Identification with a Wavelet Collocation Method for
ODEs and DAEs: 5 - 22*

*Parameter Identification with a Wavelet Collocation Method in the
Black Scholes Equation: 23 - 36*

Adapted Linear Approximation for Logarithmic Kernel Integrals: 37 - 44

*Identifying a Superposition with Trigonometric Functions by Applying a MRA
with the Shannon Wavelet: 45 - 55*

Parameter Identification with a Wavelet Collocation Method for ODEs and DAEs

M. Schuchmann and M. Rasguljajew from the Darmstadt University of Applied Sciences

Abstract

This article describes parameter identification with a wavelet collocation method which minimizes the sum of squares of residuals. In the examples we use the Shannon wavelet and a Daubechies wavelet. The parameter identification is a problem, were an ODE, PDE or DAE is given with unknown parameters. The parameters should be estimated by given measurements. Such problems appear in the chemical reaction kinetics or in n-body problems. If the problem is stiff, we need a boundary value approach. The advantage of the wavelet collocation is that we can use it for different types of problems and even for stiff problems. As an example we use the test problem of H. H. Robertson, which is based on a stiff ODE or DAE and an unstable system from H. G. Bock. In the examples we apply additionally an estimation in two steps, which leads under certain conditions to two quadratic minimization problems. For the assessment of the approximation and of the approximated parameters we use sum of squares of residuals.

Keywords: parameter identification, wavelet collocation, sinc collocation

Introduction

In the wavelet theory a scaling function ϕ is used, which belongs to a MSA (multi scale analysis). From the MSA we know, that we can construct an orthonormal basis of a closed subspace V_j, where V_j belongs to a the sequence of subspaces with the following property:

$$\ldots \subset V_{-1} \subset V_0 \subset V_1 \subset \ldots \subset L^2(\mathbb{R}),$$

$\{\phi_{j,k}(t)\}_{k \in \mathbb{Z}}$ is an orthonormal basis of V_j with $\phi_{j,k}(t) = 2^{j/2}\phi(2^j t - k)$, where ϕ ist he scaling function oft the MSA.

We use the following approximation function

$$y_j(t) := \sum_{k=k_{min}}^{k_{max}} c_k \cdot \phi_{j,k}(t) \quad, \text{with } \phi \in C^1(\mathbb{R}).$$

k_{max} and k_{min} depend on the approximation interval $[t_0, t_{end}]$ (see [35]).

Now we can approximate the solution of an initial value problem $y' = f(y,t)$ and $y(t0) = y_0$ by minimization of the following function

(1) $$Q(c) = \sum_{i=1}^{m} \left\| y_j'(t_i) - f(y_j(t_i), t_i) \right\|_2^2 + \left\| y_j(t_0) - y_0 \right\|_2^2 .$$

For $m = |k_{max} - k_{min}|$ we get an equivalent problem:

$$y_j'(t_i) = f(y_j(t_i), t_i) \text{ for } i = 1, 2, \ldots, m \text{ and } y_j(t_0) = y_0.$$

Analogous we could treat boundary conditions instead of the initial condition. This method can be even used analogous for PDEs, ODEs of higher order or ODEs, which have the Form $F(y', y, t) = 0$.

If $y' = f(y,t)$ is an ODE system, then we use the approximation function

$$y_j(t) = \left(\sum_{k=k_{min}}^{k_{max}} c_{k,1} \phi_{j,k}(t), \sum_{k=k_{min}}^{k_{max}} c_{k,2} \phi_{j,k}(t), \ldots, \sum_{k=k_{min}}^{k_{max}} c_{k,n_f} \phi_{j,k}(t) \right)^T.$$

For the i-th component of the solution y, we use - as usual, the notation y_i. We use for the i-th component of y_j the notation $y_j^{(i)}$, so that it does not lead to a confusion with the approximation y_j out of V_j and it will be always noticed in context whether it is the approximation y_j or it is the i-th component of y.

We use the collocation points t_i, with $t_i = t_0 + i \cdot h$ and

$$(2) \qquad h = \frac{t_{end} - t_0}{m} \qquad (m \geq |k_{max} - k_{min}|).$$

For the assessment of the approximation we use the value Q_a, with

$$Q_a = \sum_{i=1}^{m_a} \left\| y_j'(\tau_i) - f(y_j(\tau_i), \tau_i) \right\|_2^2 + \left\| y_j(t_0) - y_0 \right\|_2^2,$$

$\tau_i = t_0 + i \cdot h/a$, $m_a = a \cdot m$ and $a > 1$ is an integer.

If nothing is known about the solution y, many simulations have shown that a usable starting value of m is $m = |k_{max} - k_{min}|$. For this choice of m we would have for the case $n = 1$ in c together $m + 1$ coefficients, m collocation points and one initial value. In the case that y has big slopes or big curvatures, we need a bigger m. A too small m leads to a big Q_{min} or a big Q_a (see (6)).

For the parameter identification we minimize

$$(3) \qquad Q_{\alpha,\beta}(p,c) = \alpha \cdot \sum_{i=0,\ldots,m} \left\| y_j'(t_i) - f(y_j(t_i), t_i, p) \right\|_2^2 + \beta \cdot \sum_{i=1,\ldots,\tilde{m}} \left\| \tilde{m}_i - M(y_j(\hat{t}_i)) \right\|_2^2.$$

Boundary conditions or initial conditions can be used as constrains or can be considered in Q.

In the following example we minimize

$$(4) \quad Q_{\alpha,\beta}(p,c) = \alpha \cdot \sum_{i=0,\ldots,m} \left\| y_j'(t_i) - f(y_j(t_i), t_i, p) \right\|_2^2 + \beta \cdot \sum_{i=1,\ldots,\tilde{m}} \left\| \tilde{m}_i - M(y_j(\hat{t}_i)) \right\|_2^2 + \left\| y_0 - y_j(t_0) \right\|_2^2$$

with $\alpha = \beta = 1$. The (numerical calculated) minimum value of Q is Q_{min}.

A method which is a variation of the method above is the minimization in two steps:

In the first step we calculate

(5) $$Q_{0,1}(p,\hat{c}) = \min_c Q_{0,1}(p,c)$$

And in the second step we calculate

(6) $$Q_{1,0}(\hat{p},\hat{c}) = \min_p Q_{1,0}(p,\hat{c})$$

If M and f is linear in p, like in the examples in the reaction kinetics, then we must solve in both steps a quadratic problem. In that case we can apply a parameter identification with a small effort, even more if the problem is stiff.

For the assessment of the approximation we could calculate the following function value:

(6) $$Q_{\alpha,\beta,a}(\hat{p},\hat{c}) = \alpha \cdot \sum_{i=0,\ldots,m_a} \left\| y_j'(\tau_i) - f(y_j(\tau_i),\tau_i,\hat{p}) \right\|_2^2 + \beta \cdot \sum_{i=1,\ldots,\tilde{m}} \left\| \tilde{m}_i - M(y_j(\hat{t}_i)) \right\|_2^2 + \left\| y_0 - y_j(t_0) \right\|_2^2$$

with $\tau_i = t_0 + i \cdot h/a$, $m_a = a \cdot m$ and an integer $a > 1$.

In the case $\alpha = \beta = 1$ we later write in short Q_a (like Q for $Q_{1,1}$).

With the exact solution $Q_{\alpha,0,a}(p,c)$ is equal to zero. If $Q_{\alpha,\beta,a}(\hat{p},\hat{c}) \gg Q_{\alpha,\beta}(\hat{p},\hat{c})$, we need more collocation points t_i and if Q_{min} is relative big, then we need a bigger j. For this criteria we must not calculate a second minimization, if $Q_{\alpha,\beta,a}(\hat{p},\hat{c})$ is small. For $a \gg 1$ we should use $1/a \; Q_{\alpha,\beta,a}(\hat{p},\hat{c})$ instead of $Q_{\alpha,\beta,a}(\hat{p},\hat{c})$ for the comparison with $Q_{\alpha,\beta}(\hat{p},\hat{c})$.

Remarks:
1) By using the Shannon wavelet and if the measurement points $\hat{t}_i$ can be chosen freely (if this is technically possible in a practical problem), then a good choice would be $\hat{t}_{i+1} - \hat{t}_i = \Delta \hat{t} \leq 2^{-j}$, which follows from Shannon's Theorem. If y is in V_j then we would get $c_k = 2^{-j/2} y(2^{-j} \cdot k)$. That could be used for the choice of starting values for the coefficients c_k in the iteration, if $M(y) = y$.

2) In case of $M(y) = y$ we could apply a discrete wavelet transformation on the measurements, to get an information about the involved frequencies.

Example 1 (ROBER)

1966 H. H. Robertson introduced an example in the reaction kinetics ([33]), which includes a very fast reaction,. This fast reaction is responsible for the stiffness of the system. This example is a test set of the INdAM-Bari Group. E. Hairer ([22]) used the short name ROBER for this example.

The reaction schema is:

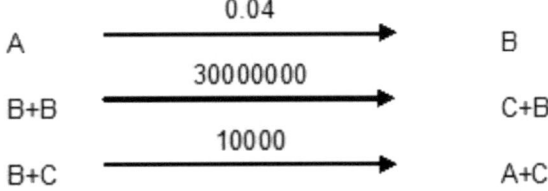

Fig. 1. Reaction schema of ROBER

Here is the resulting ODE:

$$y_1' = -p_1 y_1 + p_3 y_2 y_3$$
$$y_2' = p_1 y_1 - p_3 y_2 y_3 - p_2 y_2^2$$
$$y_3' = p_2 y_2^2 \;,$$

with $p = (0.04, 3 \cdot 10^7, 10^4)^T$. We use the starting vector $y(0) = (1, 0, 0)^T$.

With $p = (0.04, 3 \cdot 10^7, 10^4)^T$ the system cannot be solved with an explicit method. We need an implicit method or boundary value method.

H. H. Robertson 1966:
When the equations represent the behaviour of a system containing a number of fast and slow reactions, a forward integration of these equations becomes difficult.

Now we use **the wavelet collocation method** with the Shannon wavelet. We simulate measurements by using the points $\hat{t}_i = 0.1 \cdot i$, with $i = 1,...,50$, the measurement function $M(y) = y$ and the approximation interval $I = [0, 5]$, that means $t_{end} = 5$.

We set $j = 1$, $k_{max} = 25$ and $k_{min} = -5$ for y_j. k_{min} is not so small, because we start with $t_0 = 0$. The collocation points are

$$t_i = 0.05 \cdot i, \text{ with } i = 1,...,100 \;,$$

so $m = 100$.

We make two estimations of p. In the first estimation we estimate as described in two steps. In the first step we estimate c with $\hat{c}$, by setting in

$$Q_{\alpha,\beta}(p,c) = \alpha \cdot \sum_{i=1,...,100} \left\| y_j'(t_i) - f(y_j(t_i), t_i, p) \right\|^2 + \beta \cdot \sum_{i=1,...,50} \left\| \tilde{m}_v - M(y_j(\hat{t}_i)) \right\|^2 + \left\| y_j(t_0) - y_0 \right\|^2$$

the coefficients $\alpha = 0$ and $\beta = 1$ and calculate

$$Q_{0,1}(p,\hat{c}) = \min_c Q_{0,1}(p,c) \;.$$

In the second step we estimate p with

$$Q_{1,0}(\hat{p},\hat{c}) = \min_p Q_{1,0}(p,\hat{c}) \;.$$

Therefore we only have to solve a system of the normal equation twice, because the problems are quadratic (for an error analysis see [36]).

We got:

$$Q_{0,1}(p,\hat{c}) = \min_c Q_{0,1}(p,c) \approx 4.41611 \cdot 10^{-13}$$

Now we see the curves of $y_1^{(i)} - y_i$, beginning with $i = 1$ (here y is not the exact solution, but the numerical solution of the system with the exact parameter p, by using the Mathematica-function NDSolve):

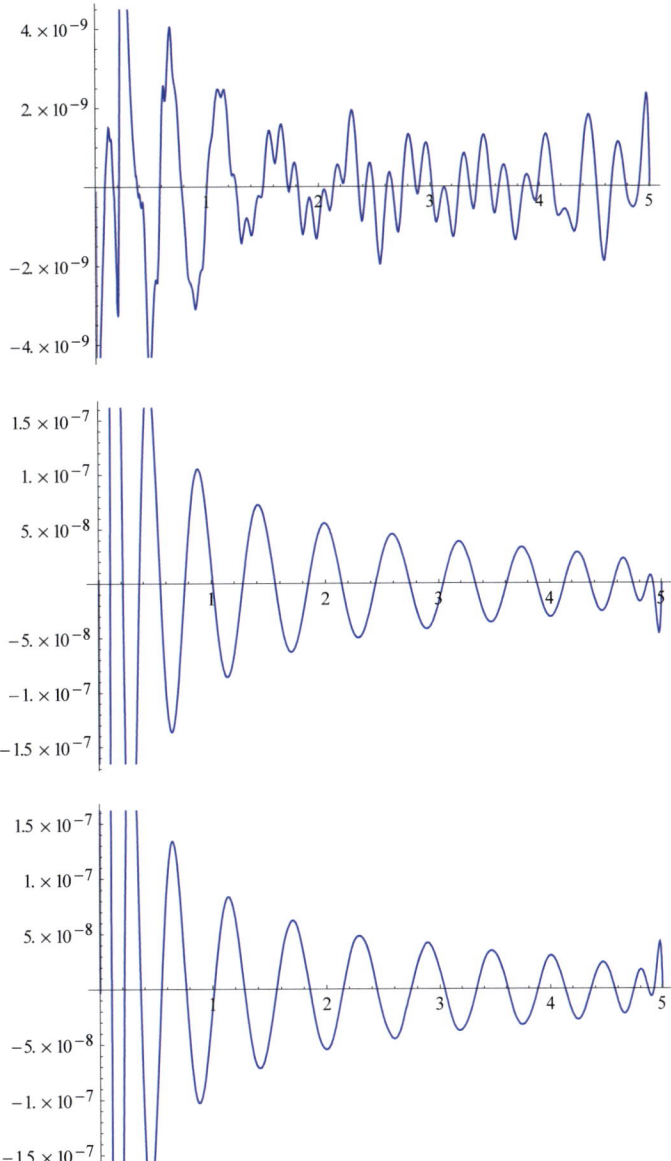

Fig. 3. Curves of $y_1^{(i)} - y_i$

Here are the graphs for $i = 2$ (y_2 in red):

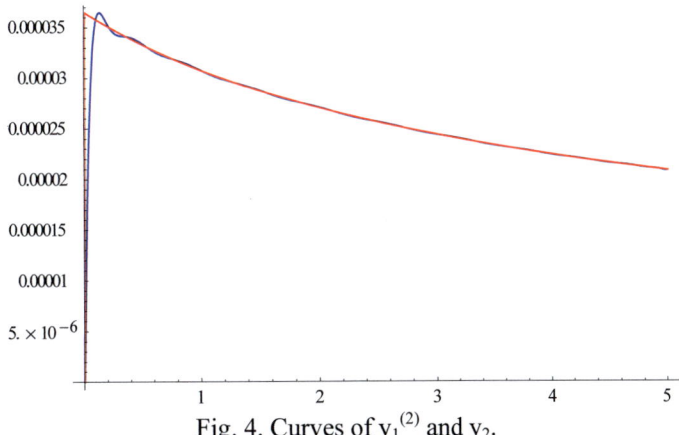

Fig. 4. Curves of $y_1^{(2)}$ and y_2.

For $y^{(2)}$ and a smaller area:

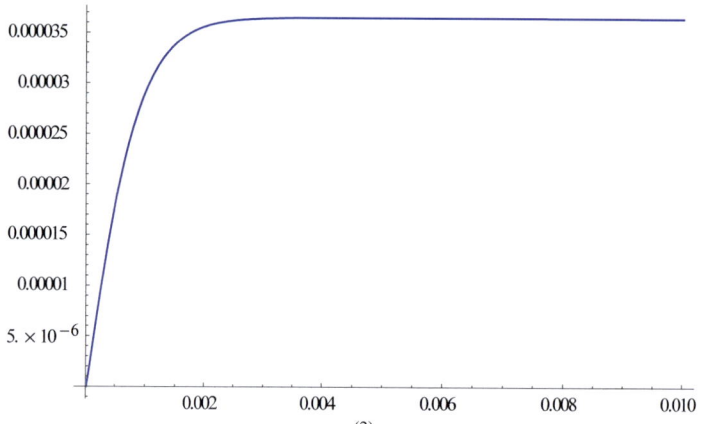

Fig. 5. Curves of $y_1^{(2)}$ at the beginning.

For the iteration the parameter has been rescaled, so instead of the system

$$y_1' = -p_1 y_1 + p_3 y_2 y_3$$
$$y_2' = p_1 y_1 - p_3 y_2 y_3 - p_2 y_2^2$$
$$y_3' = p_2 y_2^2 ,$$

with $p = (0.04, 3 \cdot 10^7, 10^4)^T$ we use the system

$$y_1' = -p_1 y_1 + 10^4 p_3 y_2 y_3$$
$$y_2' = p_1 y_1 - 10^4 p_3 y_2 y_3 - 10^7 p_2 y_2^2$$
$$y_3' = 10^7 p_2 y_2^2 ,$$

with $p = (0.04, 3, 1)^T$.

Although in step 1 in the area [0, 0.1] the approximation for $y_1^{(2)}$ was not very good but the estimation of p was not bad:

p_i	$\hat{p}_i$	$|(\hat{p}_i - p_i)/p_i|$
0.04	0.0399575	0.00106353
3	3.01548	0.00516115
1	1.00002	0.0000249715

Table 1.

Using the Daubechies wavelet of order 7 (with $k_{max} = 19$, $k_{min} = -5$ und $j = 2$) we get the following approximation in the second step:

p_i	$\hat{p}_i$	$|(\hat{p}_i - p_i)/p_i|$
0.04	0.039746	0.00634911
3	2.98796	0.00401307
1	0.999414	0.000585899

Table 2.

In many simulations with Daubechies wavelets it has been seen that wee need a bigger j but less coefficients c_i (in comparison to Shanon wavelets), because of the compact support of the Daubechies wavelets.

Now, we estimate the parameter p together with c and set $\alpha = \beta = 1$ (using the Shannon wavelet. We get:

$$\min_{c,p} Q_{1,1}(p,c) \approx 2.2264 \cdot 10^{-9}$$

For comparison: $Q_2 \approx 3.09351 \cdot 10^{-9}$.

Now we see the curves of $y_1^{(i)} - y^{(i)}$ beginning with i = 1:

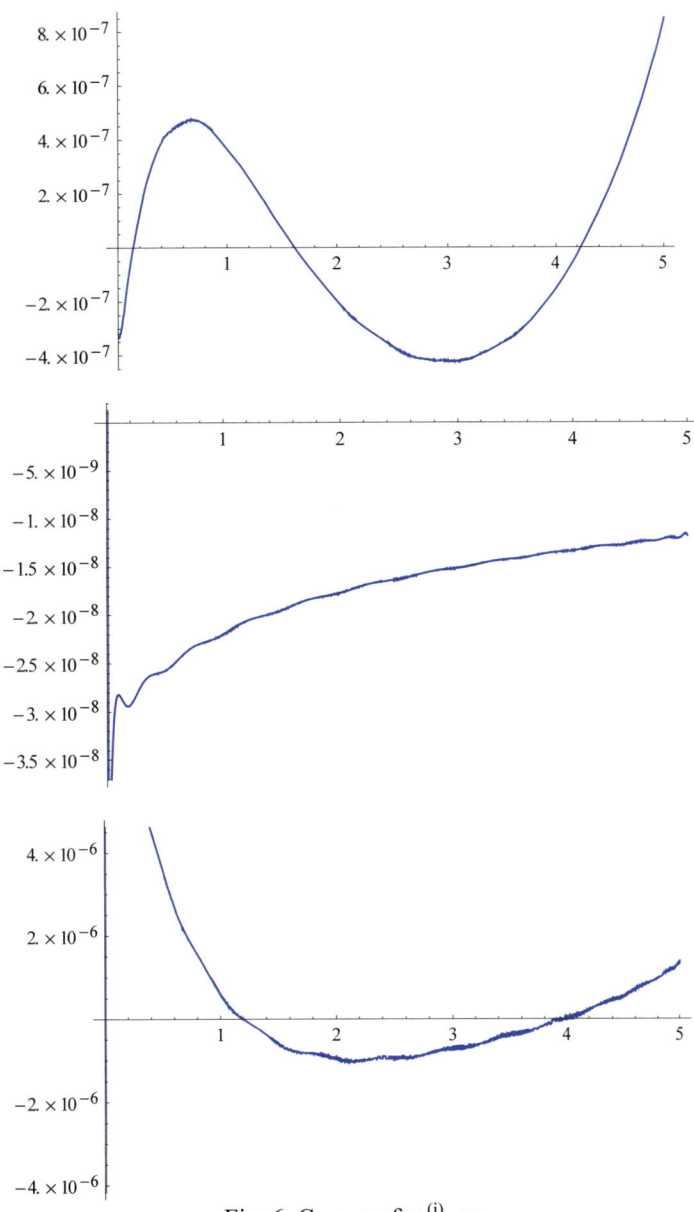

Fig. 6. Curves of $y_1^{(i)} - y_i$.

$y^{(2)}$ was approximated well (in red we see $y^{(2)}$ together with $y_1^{(2)}$):

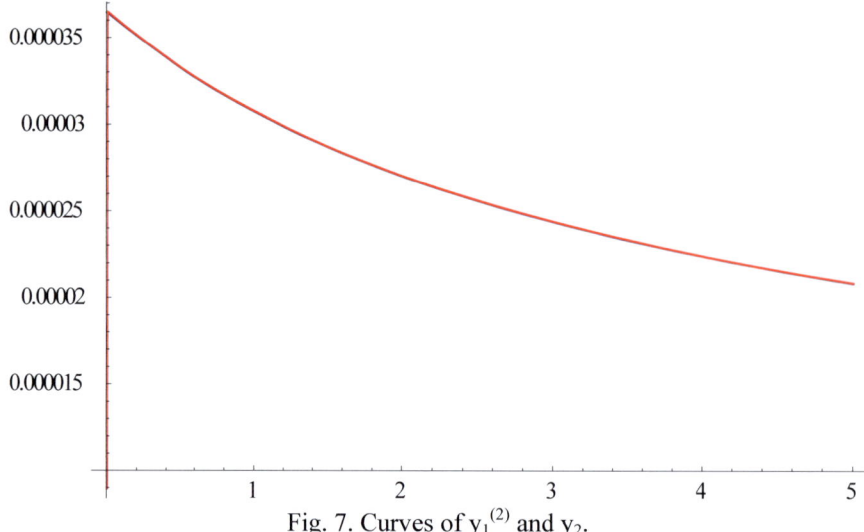

Fig. 7. Curves of $y_1^{(2)}$ and y_2.

p_i	$\hat{p}_i$	$|(\hat{p}_i - p_i)/p_i|$
0.04	0.0399879	0.000303367
3	3.00382	0.00127247
1	1.00063	0.000627453

Table 3.

For comparison, we do the same estimation with the Daubechies wavelet of order 7 (with $k_{max} = 19$, $k_{min} = -5$ und $j = 2$):

p_i	$\hat{p}_i$	$|(\hat{p}_i - p_i)/p_i|$
0.04	0.0399279	0.00180237
3	3.02962	0.00987431
1	1.00349	0.00348764

Table 4.

The method for the approximation and estimation can be applied analogous to a DAE. We can write the System in an equivalent DAE:

$y_1' = -p_1 y_1 + 10^4 p_3 y_2 y_3$
$y_2' = p_1 y_1 - 10^4 p_3 y_2 y_3 - 10^7 p_2 y_2^2$
$1 = y_1 + y_2 + y_3$

with $p = (0.04, 3, 1)^T$ and $y(0) = (1, 0, 0)^T$.

We use the same measurements and minimize

$$Q_{\alpha,\beta}(p,c) = \alpha \cdot \sum_{i=1,\ldots,100} \left\| F(y_j'(t_i), y_j(t_i), t_i) \right\|_2^2 + \beta \cdot \sum_{i=1,\ldots,50} \left\| \tilde{m}_v - M(y_j(\hat{t}_i)) \right\|_2^2 + \left\| y_j(t_0) - y_0 \right\|_2^2$$

with

$$F(y'', y', t_i) = (y_1' + p_1 y_1 - 10^4 p_3 y_2 y_3, y_2' - p_1 y_1 + 10^4 p_3 y_2 y_3 + 10^7 p_2 y_2^2, 1 - y_1 - y_2 - y_3)^T.$$

We use again $\alpha = \beta = 1$ and we get with the Shannon wavelet:

$$\min_{c,p} Q_{1,1}(p,c) \approx 2.22412 \cdot 10^{-9}$$

For comparison: $Q_2 \approx 3.80903 \cdot 10^{-9}$.

Here is the estimation:

p_i	$\hat{p}_i$	$|(\hat{p}_i - p_i)/p_i|$
0.04	0.0399879	0.000303616
3	3.00382	0.00127394
1	1.00063	0.000628092

Table 5.

Now we see the curves of $y_1^{(i)} - y^{(i)}$ beginning with $i = 1$:

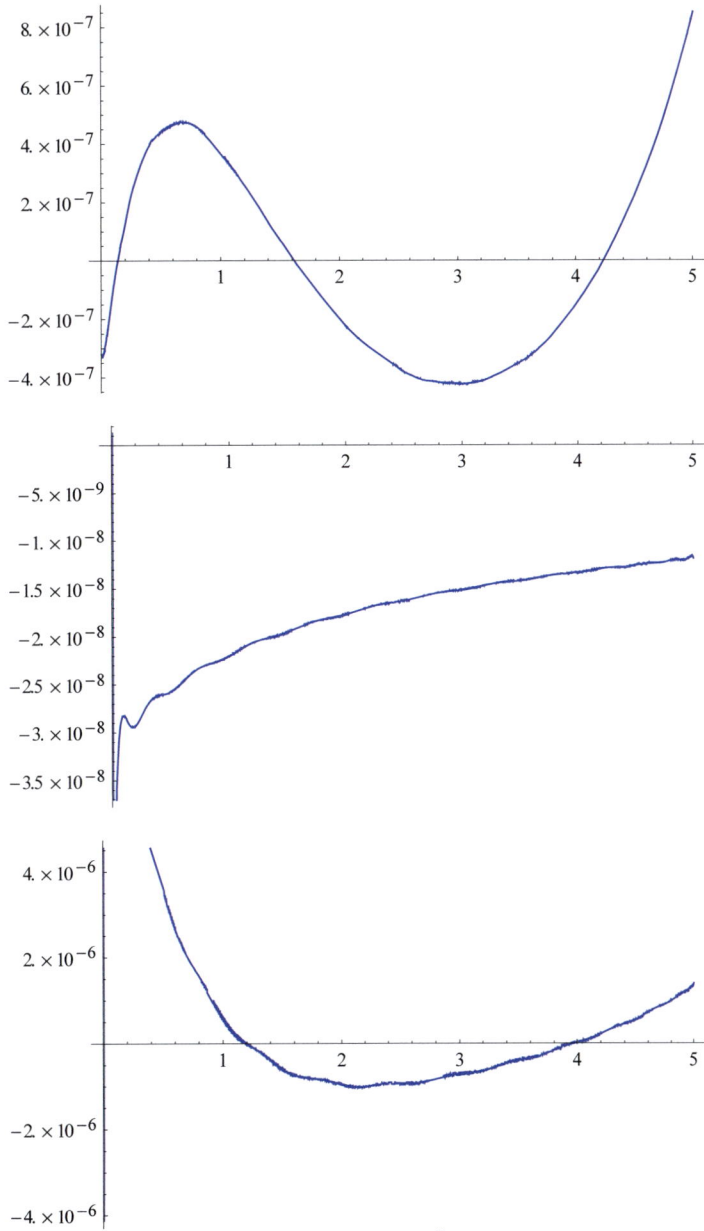

Fig. 8. Curves of $y_1^{(i)} - y_i$.

Here are the curves of the approximation functions $y_1^{(i)}$ beginning with i = 1:

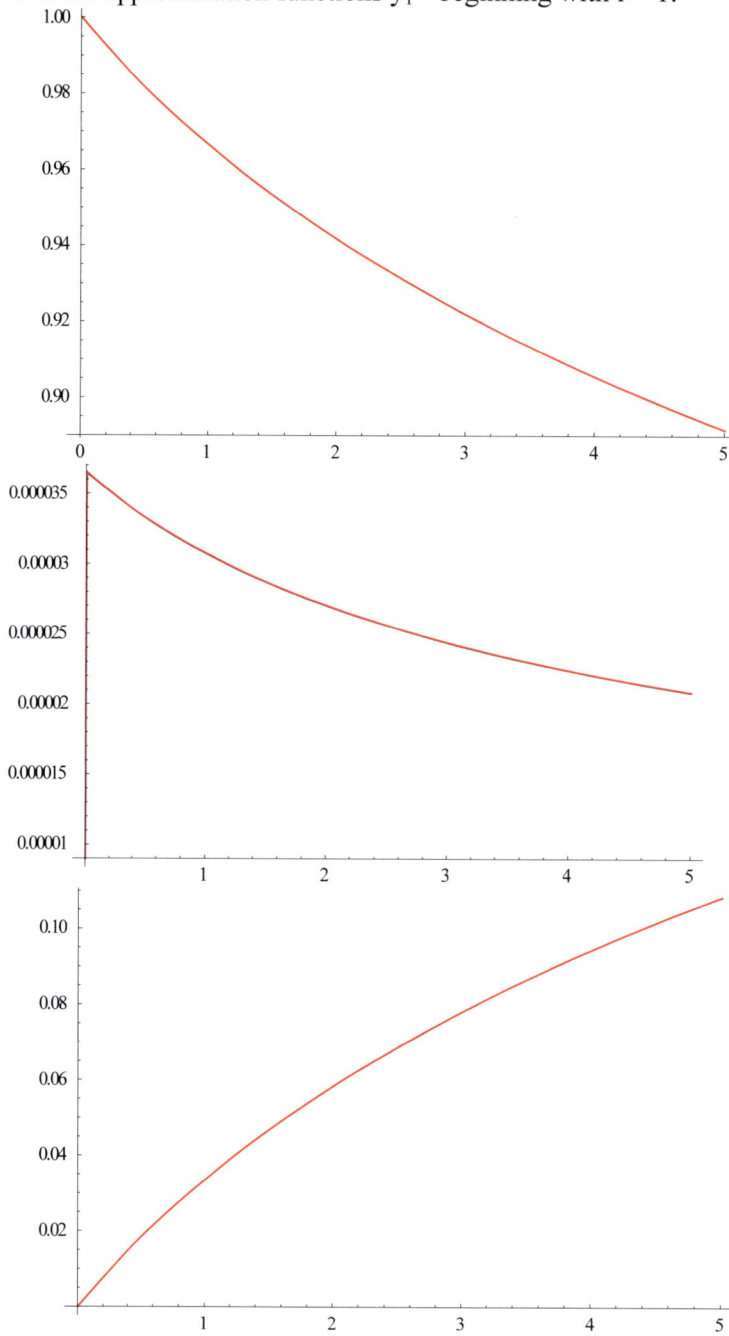

Fig. 9. Curves of $y_1^{(i)}$.

Example 2 (BOCK)

An example of a parameter identification problem with an instable System is from H. G. Bock ([12]). Here we apply the same Method of wavelet collocation.

Here is the problem:

$$y_1' = y_2$$
$$y_2' = \tau^2 \cdot y_1 - (\tau^2 + \theta^2) \cdot \sin(\theta \cdot t),$$
$$y(0) = (0, \theta)^T.$$

The solution is

$$y_1(t) = \sin(\theta \cdot t) \text{ und } y_2(t) = \theta \cdot \cos(\theta \cdot t).$$

The solution without starting value would be:

$$y_1(t) = c_1 \cdot e^{\tau \cdot t} + c_2 \cdot e^{-\tau \cdot t} + \sin(\theta \cdot t)$$

and

$$y_2(t) = \tau \cdot c_1 \cdot e^{\tau \cdot t} - \tau \cdot c_2 \cdot e^{-\tau \cdot t} + \theta \cdot \cos(\theta \cdot t).$$

When we change the value of θ only slightly, we get for bigger $|\tau|$ a strong change of the solution. For example with $\tau = 100$ and $\theta = 3$ the solution curve of y_1 looks like in the following graph:

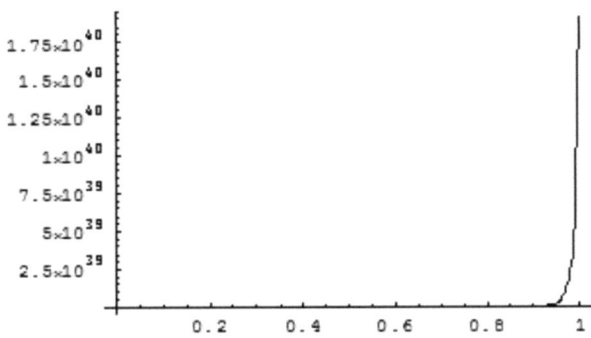

Fig. 10. Curve of y_1 with $\theta = 3$.

For $\theta = \pi$ we now see the curves of the both solutions:

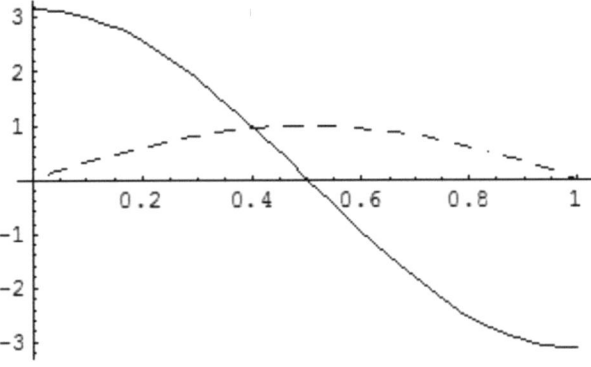

Fig. 11. Curves of y_i with $\theta = \pi$, y_1 is dashed.

Now we come to the estimation of p. We use an approximation function y_j with $j = 1$ and $k_{max} = -k_{min} = 15$. We set $\tau = 100$ and θ is the unknown parameter p: $p = \theta$. We simulated measurements with $\theta = \pi$, $M(y) = y$ and the points $\hat{t}_i = 0.1 \cdot i$, with $i = 1,...,9$ ($\hat{m} = 9$).

The approximation interval is $I = [t_0, t_{end}] = [0, 1]$ and we use the collocation points

$$t_i = 0.05 \cdot i, \text{ with } i = 1,...,20$$

(m = 20) and set $\alpha = \beta = 1$ and use the constrain $y_j(0) = y(0)$ and the function:

$$Q_{\alpha,\beta}(p,c) = \alpha \cdot \sum_{i=0,\ldots,m} \left\| y_j'(t_i) - f(y_j(t_i), t_i, p) \right\|_2^2 + \beta \cdot \sum_{i=1,\ldots,\hat{m}} \left\| \tilde{m}_i - M(y_j(\hat{t}_i)) \right\|_2^2$$

The minimal value of Q was: $Q_{min} \approx 3.02674 \cdot 10^{-23}$

For a comparison: $Q_2 \approx 7.32641 \cdot 10^{-20}$

We've got the estimator $\hat{p} = 3.1415926535896475$ ($|\hat{p} - \pi| \approx 1.45661 \cdot 10^{-13}$).

Here are the curves of $y_1^{(i)} - y_i$ beginning with i = 1:

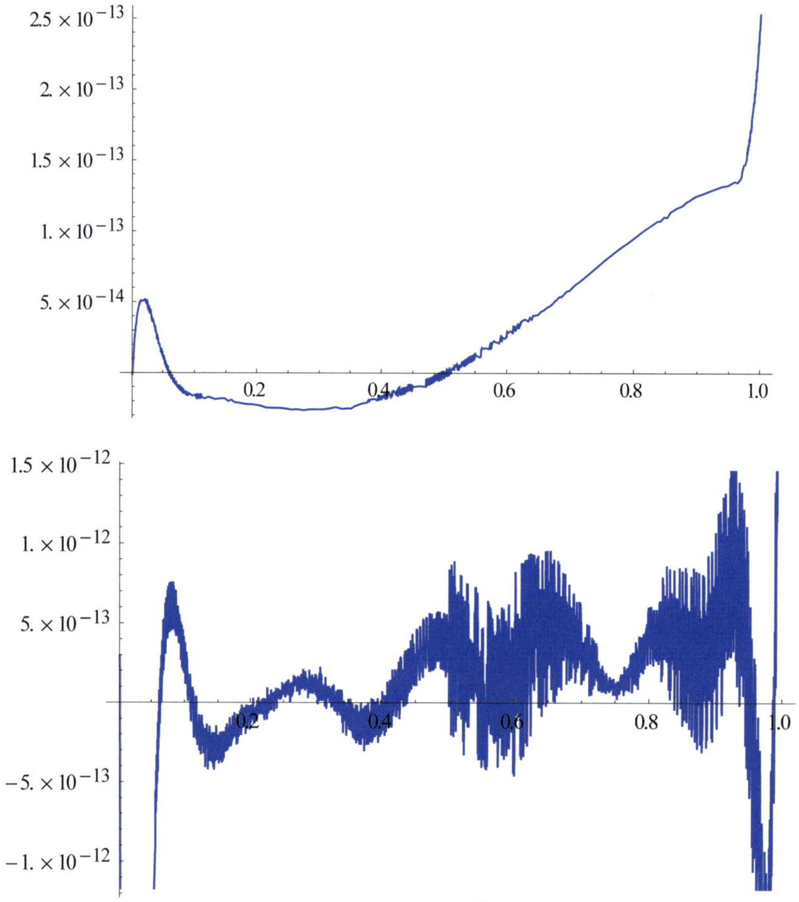

Fig. 12. Curves of $y_1^{(i)}$ - y_i.

Now we see that there is a correlation between Q_{min} and the error of the estimation and between Q_{min} and the sum of squares of the approximation error.

We estimated the parameter p by minimizing $Q_{\alpha,\beta}$ without constraints with different j and k_{max} (j = -1, 0, 1, 2, k_{max} = 15, 20, 25). With more iteration steps we could get even smaller Q_{min}, but we want to show the correlation between Q_{min} and the errors. With Q_{min} or better with Q_a we can check if we used a too small j or not enough collocation points.

Because of the term $\sin(\theta \cdot t)$ in the solution with $\theta = \pi$ the parameter j should not be less than zero, if we use the Shannon wavelet. The reason is: With the Shannon theorem we know, that y is in V_j if the support of the Fourier transform Y is a subset of $[-2^j \cdot \pi, 2^j \cdot \pi]$. So for the function $h(t) = \sin(at)$ the parameter a should be in $[-2^j \cdot \pi, 2^j \cdot \pi]$.

Here is a table with the results of the estimation for different j and k_{max}:

j	k_{max}	Q_{min}	$\hat{p}$	$|\hat{p} - \pi|$
-1	15	31.370117	0.8081795	2.333413
-1	20	29.996680	1.1156355	2.025957
-1	25	31.305504	0.8332533	2.308339
0	15	5.12784E-07	3.141578325	1.43284E-05
0	20	2.04753E-07	3.141583363	9.29025E-06
0	25	1.70835E-07	3.141584148	8.50521E-06
1	15	4.19137E-07	3.141584247	8.40653E-06
1	20	3.44747E-07	3.141584891	7.76269E-06
1	25	3.34268E-07	3.141584929	7.72466E-06
2	15	5.42453E-06	3.141540568	5.20854E-05
2	20	7.69676E-06	3.141532863	5.97911E-05
2	25	1.7566E-07	3.141583311	9.34276E-06

Table 6.

We see that Q_{min} seems to correlate with the error of the estimation. For that reason we apply a linear regression analysis to the points $(-\ln(Q_{min}), -\ln(|\hat{p} - \pi|))$ which have been calculated with different j and k_{max}.

Here is the output:

	Estimate	SE	t-Stat	p-Value
1	1.53532	0.246562	6.22691	0.0000979527
x	0.642696	0.0136946	46.9307	$4.65061 \cdot 10^{-13}$

Table 7.

R-Squared = 0.997948

The linear regression fits very well.

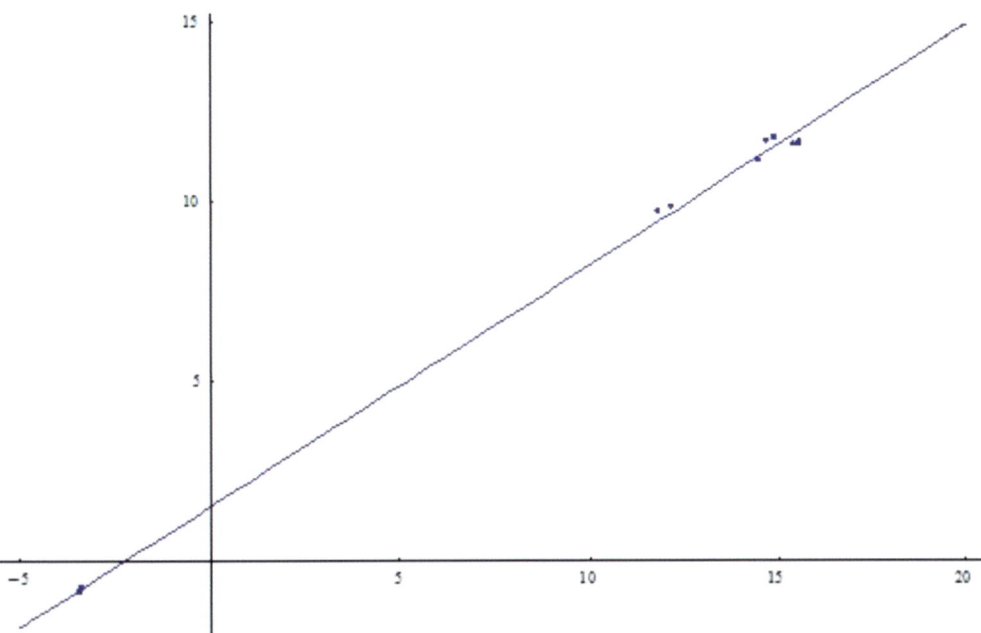

Fig. 13. $\ln(Q_{min})$ vs. $-\ln(|\hat{p} - \pi|)$ and regression with linear regression line.

Now we see a second correlation between Q_{min} and the squared error

$$sse = \sum_{i=0}^{100} (y(t_0 + i \cdot h_0) - y_j(t_0 + i \cdot h_0))^2 \quad \text{with} \quad h_0 = (t_{end} - t_0)/100 = 0.01,$$

if we look at the table:

j	k_{max}	Q_{min}	sse
-1	15	31.37011658	281.964
-1	20	29.9966803	271.717
-1	25	31.30550382	281.658
0	15	5.12784E-07	1.45331E-06
0	20	2.04753E-07	5.81371E-07
0	25	1.70835E-07	4.86926E-07
1	15	4.19137E-07	1.44924E-06
1	20	3.44747E-07	1.19198E-06
1	25	3.34268E-07	1.15528E-06
2	15	5.42453E-06	2.51805E-05
2	20	7.69676E-06	3.47701E-05
2	25	1.7566E-07	7.96269E-07

Table 8.

We now apply a linear regression analysis on the points $(-\ln(Q_{min}), -\ln(sse))$ which have been calculated with the different j and k_{max}.

	Estimate	SE	t-Stat	p-Value
1	-2.02816	0.0706197	-28.7194	$6.1004 \cdot 10^{-11}$
x	1.05317	0.00559448	188.252	$4.39696 \cdot 10^{-19}$

Table 9.

R-Squared = 0.999718

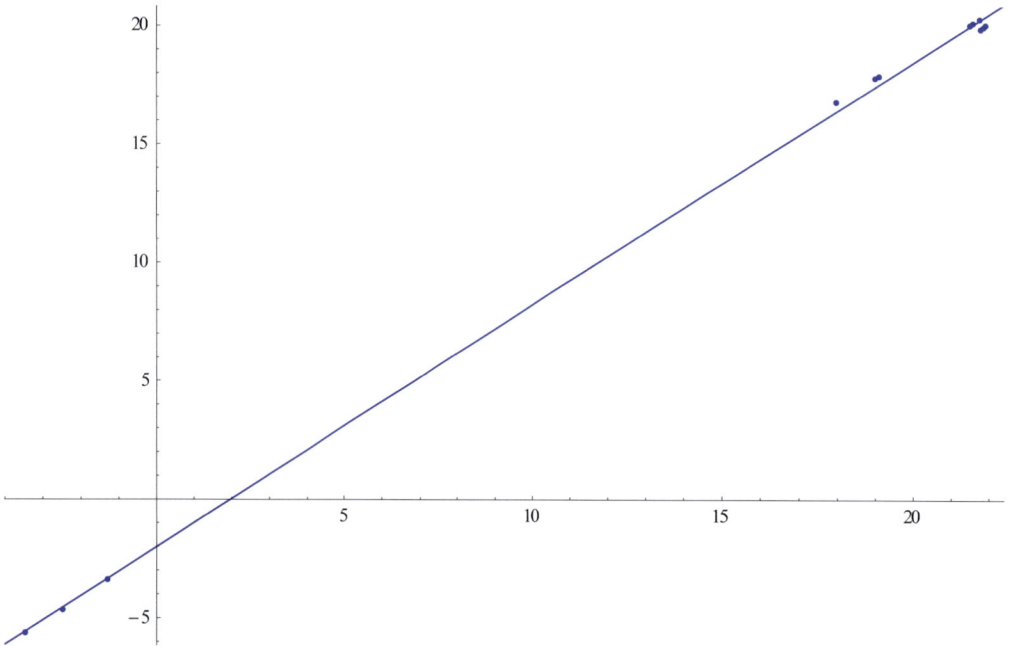

Fig. 14. $-\ln(Q_{min})$ vs. $-\ln(sse)$ and regression with linear regression line.

At least we estimate the parameter p in two steps, so we must solve only two times a system of the normal equation. We use

$$Q_{\alpha,\beta}(p,c) = \alpha \cdot \sum_{i=1,\ldots,20} \left\| y_j'(t_i) - f(y_j(t_i), t_i, p) \right\|^2 + \beta \cdot \sum_{i=1,\ldots,9} \left\| \tilde{m}_v - M(y_j(\hat{t}_i)) \right\|^2$$

and estimate at first c with $\alpha = 0$ and $\beta = 1$:

$$Q_{0,1}(p, \hat{c}) = \min_c Q_{0,1}(p, c)$$

Then we estimate p:

$$Q_{1,0}(\hat{p}, \hat{c}) = \min_p Q_{1,0}(p, \hat{c})$$

In the first step, we get: $Q_{0,1}(p, \hat{c}) = \min_c Q_{0,1}(p, c) \approx 2.78704 \cdot 10^{-30}$.

Here are the curves of $y_j - y$:

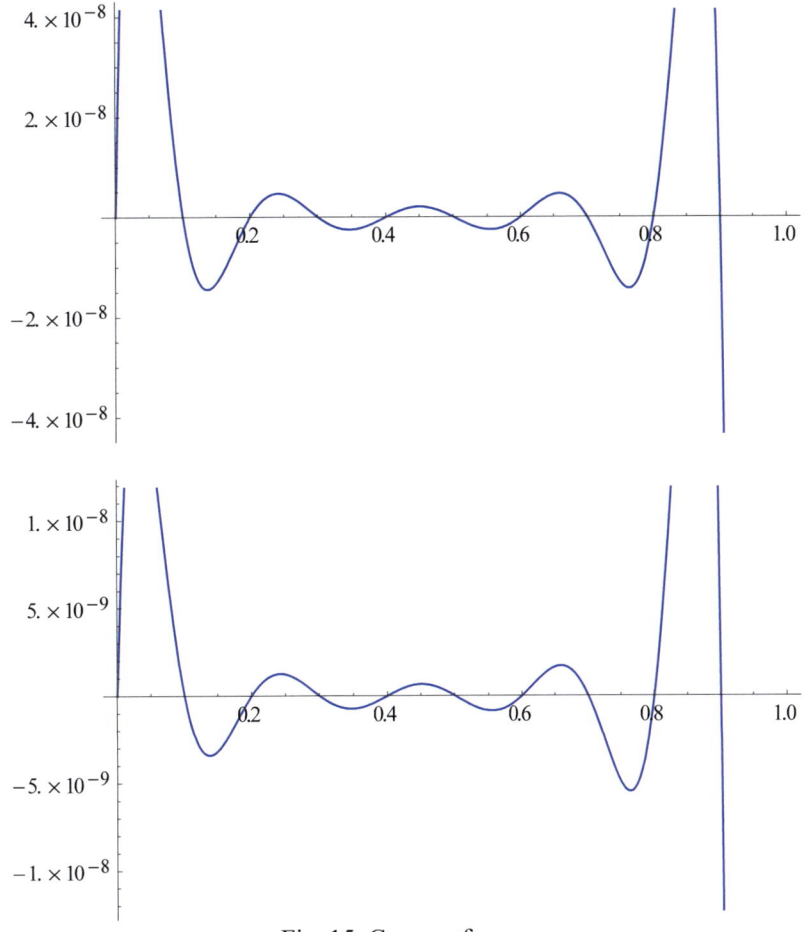

Fig. 15. Curves of $y_j - y$.

In the second step, we estimate p with

$$Q_{1,0}(\hat{p}, \hat{c}) = \min_p Q_{1,0}(p, \hat{c})$$

and we get the following error $|\hat{p} - \pi| \approx 1.79995 \cdot 10^{-6}$.

References

[1] Abdella, K. (2012). Numerical Solution of two-point boundary value problems using Sinc interpolation. *Proceedings of the American Conference on Applied Mathematics (American-Math '12): Applied Mathematics in Electrical and Computer Engineering*

[2] Ascher, U. A. Mattheij, R. M. M. Russell, R. D. (1988). Numerical Solution of Boundary Value Problems for ODEs. *Prentice Hall (Series in Computational Mathematics)*

[3] Ascher, U. Christiansen, J. Russell, R. (1981). Collocation Software for Boundary Value ODEs. *ACM Trans. Math. Software*

[4] Bauer, I. Bock, H. G. Körkel, S. Schlöder, J. P. (2000). Numerical Methods for Optimum Experimental Design in DAE Systems. *Journal of Computational and Applied Mathematics*

[5] Biegler, L. T. (1984). Solution of Dynamic Optimization Problems by Successive Quadratic Programming and Orthogonal Collocation. *Comput. Chem. Engng.*

[6] Biegler, L. T. (2002). Advances in Simultaneous Strategies for Dynamic Systems. *Chemical Engineering Sciences*

[7] Binder, T. Blank, L. Bock, H. G. Burlisch, R. et. al. (2001). Introduction to Model Based Optimization of Chemical Processes on Moving Horizons. *In: Online Optimization of Large Scale Systems" Springer Verlag*

[8] Bertoluzza S. (2006). Adaptive wavelet collocation method for the solution of Burgers equation. *Transport Theory and Statistical Physics, 25: 3-5*

[9] Bock, H. G. (1978). Numerical Solution of Nonlinear Multipoint Boundary Value Problems with Application to Optimal Control. *Z. Angew. Math. Mech.*

[10] Bock, H. G. (1981). Numerical Treatment of Inverse Problems in Chemical Reaction Systems. *Springer Series in Chem. Phys.*

[11] Bock, H. G. (1983). Recent Advances in Parameter Identification Techniques for ODE. *Numerical Treatment of Inverse Problems in Differential and Integral Equations (P. Deuflhard und E. Hairer Ed.) Birkhäuser*

[12] Bock, H. G. (1987). Randwertproblemmethoden zur Parameteridentifizierung in Systemen nichtlinearer Differentialgleichungen. *Bonner Mathematische Schriften Nr. 183*

[13] Bock, H. G. Eich. E. Schlöder, J. P. (1988) . Numerical Solution of Constrained Least Squares Boundary Value Problems in DAE. *in "Numerical Treatment of Differential Equations" BG Teubner Leipzig*

[14] Bock, H. G. Schlöder, J. P. (1987) . Recent Progress in the Development of Algorithms and Software for Large Scale Parameter Estimation Problems in Chemical Reaktion Systems. *Preprint 441, Research Report (123/256) Univerity Bonn/ Univerity Heidelberg*

[15] Carlson, T. S. Dockery, J. Lund, J. (1997). A Sinc-Collocation Method for Initial Value Problems. *Mathematics and Computation, Vol. 66, No. 217*

[16] Deuflhard, P. (1974). A Modified Newton Method for the Solution of Ill-Conditioned Systems of Nonlinear Equations with Application to Multiple Shooting. *Numer. Math. 22, 289*

[17] Deuflhard, P. Hairer, E. (1983). Numerical Treatment of Inverse Problems in Differential and Integral Equations. *Band 2 Birkhäuser*

[18] Diehl, M. Bock, H. G. Schlöder, J. P. Findeisen, R. Nagg, Z. Allgöwer, F. (2002). Real – Time Optimization and Nonlinear Model Predictive Control of Processes Governed by DAEs. *Journal of process control Vol.12, No. 4*

[19] Ebert, K. H. Deuflhard, P. Jaeger, W. (Ed.) (1981). Modelling of Chemical Reaction Systems. *Springer Ser. Chem. Phys. 18*

[20] Fogler, H. S. (1999). Elements of Chemical Reaction Engineering. *3. Auflage Prentice Hall International Series in the Physical and Chemical Engineering Sciences*

[21] Guay, M. McLean, D. D. (1995). Optimization and Sensitivity Analysis for Multiresponse Parameter Estimation in Systems of ODEs. *Computers and Chemical Engineering*

[22] Hairer, E. Wanner, G. (1993). Vol. 1 : Nonstiff Problems. *Springer 2. Auflage*

[23] Hairer, E. Wanner, G. (1996). Vol. 2 : Stiff and Differential-Algebraic Problems. *Springer 2. Auflage*

[24] Kameswaran, S. Biegler, L. T. (2006). Simultaneous Dynamic Optimization Strategies: Recent Advances and Challenges. *Computers and Chemical Engineering*

[25] Kiehl, M. (1998). Sensitivity Analysis of Stiff and Non-Stiff Initial-Value Problems. *International Series of Numerical Mathematics Birkhäuser Verlag Basel*

[26] Leineweber, D. B. Bock, H. G. Schlöder, J. P. et al. (1997). A Boundary Value Approach to the Optimization of Chemical Process Described by DAE Models. *Computers and Chemical Engineering*

[27] Lohmann, T. Bock, H. G. Schlöder, J. P. (1992). Numerical Methods for Parameter Estimation and Optimal Experiment Design in Chemical Reaction Systems. *Ind. Eng. Chem. Res.*

[28] Lohmann, T. W. (1999). Modelling and Parameter Estimation of Reaction Kinetics and Coal Pyrolysis. *Journal of Inverse and Ill Posed Problems Vol. 7*

[29] Mei, S.-L. Lv, H.-L. Ma, Q. (2008). Construction of Interval Wavelet Based on Restricted Variational Principle and Its Application for Solving Differential Equations. *Hindawi Publishing Corporation Mathematical Problems in Engineering*

[30] Nowak, U. Deuflhard, P. (1985). Numerical Identification of Selected Rate Constants in Large Chemical Reaction Systems. *Appl. Num. Math. 1*

[31] Nowak, U. Deuflhard, P. (1986). Efficient Numerical Simulation and Identification of Large Chemical Reaction Systems. *Preprint SC-86-1 des Konrad Zuse-Zentrums für Informationstechnik Berlin (ZIB)*

[32] Nurmuhammada, A. Muhammada, M., Moria, M. Sugiharab, M. (2005). Double exponential transformation in the Sinc-collocation method for a boundary value problem with fourth-order ordinary differential equation. *Journal of Computational and Applied Mathematics*

[33] Qian, L. (2002). On the Regularized Whittaker-Koltel'nikov-Shannon Sampling Theorem. *Proceedings of the Amarican Mathematical Society, Vol. 131, No. 4*

[34] Robertson, H. H. (1975). Some Properties of Algorithms for Stiff Differential Equations. *J. Inst. Math. Applics.*

[35] Russell, R. D. Christiansen, J. (1979). A Collocation Solver for Mixed Order Systems of Boundary Value Problems. *Mathematics of Computation*

[36] M. Schuchmann, M. Rasguljajew; (2013). An Approach for a Parameter Estimation with a Wavelet Collocation Method. *Journal of Approximation Theory and Applied Mathematics, Vol. 1.*

[37] Schuchmann, M. (2008). Parameteridentifikation dynamischer Systeme auf günstigen Pfaden. *DAV*

[38] Schulz, V. H. Bock, H. G. Steinbach, M. C. (1998) . Exploiting Invariants in the Numerical Solution of Multipoint Boundary Value Problems for DAE. *Siam Journals Online*

[39] Scitovski, R. Jukic, D. (1996). A Method for Solving the Parameter Identification Problem for ODEs of the second Order. *Applied Mathematics and Computation - Elsevier*

[40] Unser, M. (1996). Vanishing moments and the approximation power of wavelet expansions. *Proceedings of the 1996 IEEE International Conference on Image Processing*

[41] Unser, M. Blu, T. (1998). Comparison of Wavelets from the Point of View of their Approximation Error. *Proc. Of SPIE Vol. 3458, Wavelet Applications in Signal and Image Processing*

[42] Vuduc, R. (2000). A Wavelet Collocation Method for Solving PDEs. *J. Comp. Phys.*

Parameter Identification with a Wavelet Collocation Method in the Black Scholes Equation

M. Schuchmann and M. Rasguljajew from the Darmstadt University of Applied Sciences

Abstract

In this article we describe a parameter identification method for the Black Scholes equation, solving an inverse problem in the financial mathematics. We use a wavelet collocation method and we show in a simulation that the error of the parameter estimation and of the approximation correlates with a sum of squares of the residuals. So we can assess the approximation function and the estimated parameters. This method can be applied analogous for PDEs of higher order. In an example we use the Shannon wavelet.

Introduction of the collocation method

As an example we want to solve numerically a PDE (with boundary conditions) of first order

$$F(u(x,y), u_x(x,y), u_y(x,y), x, y) = 0,$$

$$u(x,0) = h(x)$$

on the area $D = [a_1, b_1] \times [a_2, b_2]$. Generally we use a scaling function $\phi \in C^r$ (if the order of the PDE is less or equal r). With that scaling function we can construct a two dimensional scaling function with

$$\phi(x,y) = \phi(x) \cdot \phi(y)$$

and we get the bases elements of V_j with

$$\phi_{j,k_1,k_2}(x,y) = 2^j \phi(2^j x - k_1, 2^j y - k_2) \text{ and } k_1, k_2 \in \mathbb{Z}.$$

With those basis elements we construct an approximation function (for easier notation we don't use the index j in g):

$$g(x,y) = \sum_{k_1=n_u}^{n_o} \sum_{k_2=m_u}^{m_o} \phi_{j,k_1,k_2}(x,y) \cdot c_{k_1,k_2}$$

Now we have $n_k = (n_o - n_u + 1) \cdot (m_o - m_u + 1)$ unknown coefficients c_{k_1,k_2}.

To get an approximation of the solution u we can solve the following equation:

(1a) $\quad F(g(x_i, y_l), g_x(x_i, y_l), g_y(x_i, y_l), x_i, y_l) = 0, \quad i = 0, \ldots, n_1, l = 0, \ldots, n_2$

(1b) $\quad g(z_e, 0) = h(z_e), \quad e = 0, \ldots, n_3$

With $(x_i, y_l) \in D$, $z_e \in [a_1, b_1]$, $n_k = (n_1+1) \cdot (n_2+1) + n_3 + 1$ and the collocation points $x_i \neq x_e$, $y_i \neq y_e$, $z_i \neq z_e$ for $i \neq e$.

Instead of the equations above we solve a minimization problem:

$$\min Q(c)$$

with

$$Q(c) = \sum_i \sum_l F(g(x_i,y_l), g_x(x_i,y_l), g_y(x_i,y_l), x_i, y_l)^2 + \sum_e (g(z_e,0) - h(z_e))^2$$

This has the advantage that we can use more collocation points.

By using the same collocation points as in equations (1) the minimization problem is equivalent to the former problem. If we use more collocation points in the minimum problem the equations (1) are just approximately fulfilled. But with a good approximation function g the minimum of Q is very small. If F is linear then we have a quadratic minimization problem.

We use the collocation points:

$$x_i = a_1 + i \frac{b_1 - a_1}{n_1}, \quad i = 0, \ldots, n_1;$$

$$y_l = a_2 + l \frac{b_2 - a_2}{n_2}, \quad l = 0, \ldots, n_2;$$

$$z_e = a_1 + e \frac{b_1 - a_1}{n_3}, \quad e = 0, \ldots, n_3.$$

Remark:
A possible choice of the limits of the summation n_o, n_u, m_o and m_u is such that

(2) $$\phi_{j,k_1,k_2}(x,y) = 0$$

for $(x,y) \in D$ and $k_1 > n_o, k_1 < n_u, k_2 > m_o, k_2 < m_u$. If ϕ does not have a compact support or to write generally we can replace (2) with

$$|\phi_{j,k_1,k_2}(x,y)| \leq \varepsilon$$

with a suitable $\varepsilon > 0$. So the limits of summation depend on the approximation area D. In the example we will see that although the scaling function of the Shannon wavelet does not have a compact support we do not need many base elements for a good approximation.

For the assessment of an approximation we compare $Q_{min} = \min Q(c)$ with

$$Q_a = \sum_i \sum_l F(g(\tilde{x}_i,\tilde{y}_l), g_x(\tilde{x}_i,\tilde{y}_l), g_y(\tilde{x}_i,\tilde{y}_l), \tilde{x}_i, \tilde{y}_l)^2 + \sum_e (g(\tilde{z}_e,0) - h(\tilde{z}_e))^2,$$

$$\tilde{x}_i = a_1 + i \frac{b_1 - a_1}{a \cdot n_1}, \quad i = 0,\ldots,n_1;$$

$$\tilde{y}_l = a_2 + l \frac{b_2 - a_2}{a \cdot n_2}, \quad l = 0,\ldots,n_2;$$

$$\tilde{z}_e = a_1 + e \frac{b_1 - a_1}{a \cdot n_3}, \quad e = 0,\ldots,n_3.$$

with an integer $a > 0$. Here g is the approximation function with c calculated by the minimization problem. In many simulations we got with $a = 2$ very good results like in example 1. The advantage of this method is that for Q_a we don't have to calculate a second estimation because we get not only points but a whole approximation function g. If Q_a is too big and Q_{min} is very small we need more collocation points.

Analogous we can solve an PDE of higher order numerically with the described method, for example

$$F(u, u_x, u_y, u_{yy}, x, y) = 0,$$

$$u(x,0) = h_1(x) \text{ and } u_y(x,0) = h_2(x),$$

if we minimize

$$Q(c) = \sum_i \sum_l F(g(x_i,y_l), g_x(x_i,y_l), g_y(x_i,y_l), g_{yy}(x_i,y_l), x_i, y_l)^2$$
$$+ \sum_e (g_y(z_e,0) - x)^2 + \sum_e (g(z_e,0) - 0)^2$$

We will now apply the same method to a parameter identification problem.

Parameter Estimation and Assessment of the Approximation

We use as an example the following parameter identification problem with two parameters p_1 and p_2:

$$F(v, v_s, v_t, v_{ss}, s, t) = v_t + 1/2 \cdot \sigma^2 s^2 \cdot v_{ss} + r \cdot s \cdot v_s - r \cdot v = 0,$$

$$v(s,T) = \underbrace{1_{[K,\infty)}(s) \cdot (s - K)}_{:= R(s)} .$$

Additionally we need measurements $\tilde{m}_{i,l}$ from v at the points $(\hat{s}_i, \hat{t}_l) \in [0, 2] \times [0, T]$ with $T = 2$. We set $K = 1$ and for the simulation we use $r = 0.01$ and $\sigma = 0.8$. As measurement points we choose $\hat{s}_i = i \cdot 1/10$, $i = 0, \ldots, 19$, $\hat{t}_i = i \cdot 1/10$ $i = 0, \ldots, 19$ and we use the same collocation points $s_i = i \cdot 1/10$, $i = 0, \ldots, 19$, $t_i = i \cdot 1/10$ $i = 0, \ldots, 19$. The unknown parameter vector is $p = (\sigma, r)^T$.

The exact solution of the PDE with boundary conditions:

$$v(s,t) = s \cdot \Phi\left(\frac{\ln(s/K) + (r + 1/2\sigma^2)(T-t)}{\sigma\sqrt{T-t}}\right) - K \cdot e^{-r(T-t)} \cdot \Phi\left(\frac{\ln(s/K) + (r + 1/2\sigma^2)(T-t)}{\sigma\sqrt{T-t}} - \sigma \cdot \sqrt{T-t}\right)$$

Φ is the CDF of the standard normal distribution function.

Now Q depends on the two vectors c and p we minimize $Q(c, p)$:

$$\min_{c,p} Q(c,p)$$

with

$$Q(c,p) = \sum_i \sum_l F(g(s_i,t_l), g_s(s_i,t_l), g_t(s_i,t_l), g_{ss}(s_i,t_l), s_i, t_l)^2$$
$$+ \sum_e (g(s_e, 0) - R(s_e))^2 + \sum_i \sum_l (\tilde{m}_{i,l} - g(\hat{s}_i, \hat{t}_l))^2$$

Using this method we even could identify parameters in the boundary conditions.

Now we calculate parameter estimations with $j = 0, 1, 2$ and $n_0 = m_0 = 2k_{max}$, $n_u = m_u = -k_{max}$ and $k_{max} = 3, 4, 5, 6$. We use the scaling function from the Shannon wavelet, so $\phi \in C^\infty$.

The estimated parameter we call $\hat{p}$. Here is the table of the results:

j	k_{max}	Q_{min}	Q_2	sse	$\|p - \hat{p}\|$
0	3	0.00429676	0.00950414	0.00792356	0.000248112
0	4	0.00122589	0.00350103	0.000717968	0.000570588
0	5	0.000927176	0.00388506	0.000976833	0.00217147
0	6	0.000882008	0.0033812	0.00110282	0.00250282
1	3	0.00532685	0.0132514	0.00908193	0.037558
1	4	0.0010275	0.00378688	0.000802606	0.000866278
1	5	0.000449969	0.00513302	0.0027669	0.000238822
1	6	0.000333985	0.344597	0.0575867	0.000571417
2	3	82.4355	111.977	1991.87	0.801672
2	4	1.72551	5.70822	12.184	0.817138
2	5	0.00107297	0.022765	0.00360529	0.0096349
2	6	0.000172314	0.665934	0.408905	0.000209513

Q_{min} was calculated numerically by using the Mathematica function FindMinimum. *sse* is the error sum of squares calculated with:

$$sse = \sum_{i=0}^{99} \sum_{j=0}^{99} (v(i/50, j/50) - g(i/50, j/50))^2$$

Here we see a correlation between Q_{min} and *sse*, Q_2 and *sse* and between Q_2 (or Q_{min}) and *sse*. In many simulations we saw that Q_2 is a better criterion to assess the estimation and the approximation because in Q_2 more points than the collocation points (with which we minimized Q) are considered. In many simulations we saw that in a bad approximation Q_{min} can be small and Q_2 (or general Q_a) is relative big because the exact solution fulfils the PDE and the boundary value conditions at any point of the approximation area.

If we use a bigger j then we need a bigger k_{max}, too, because with rising j we compress the bases functions ϕ_{j,k_1,k_2} .

In this example we have (as we did in other simulations) a linear dependency between $ln(Q_2)$ and $ln(sse)$.

This is what we see if we take a look at the regression table:

$ln(Q_2)$ vs. $ln(sse)$:

	Estimate	SE	TStat	PValue
1	0.295856	0.341634	0.866001	0.406787
x	1.30542	0.0771474	16.9211	1.09262×10^{-8}

, RSquared → 0.966253

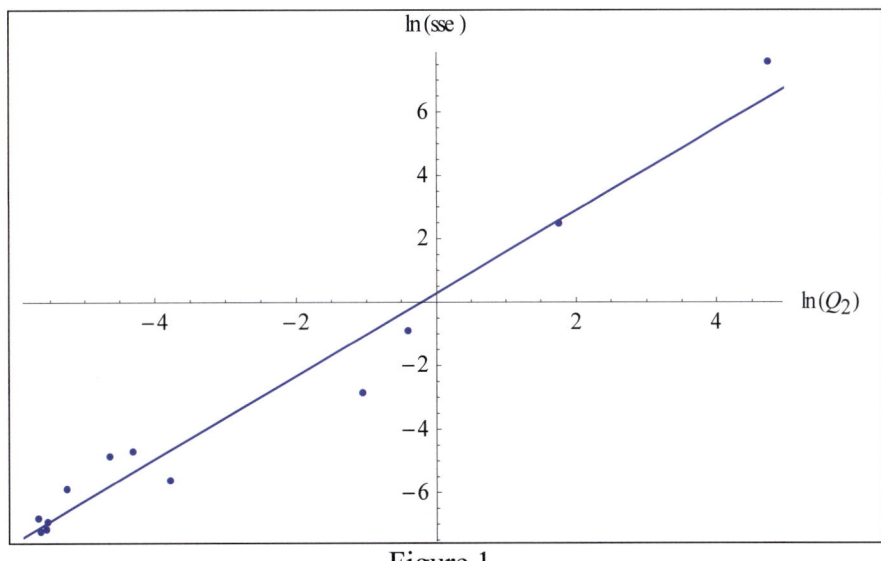

Figure 1

Here is the graph of $v - g$ for $j = 0$ and $k_{max} = 6$ (with the small Q_2 and sse):

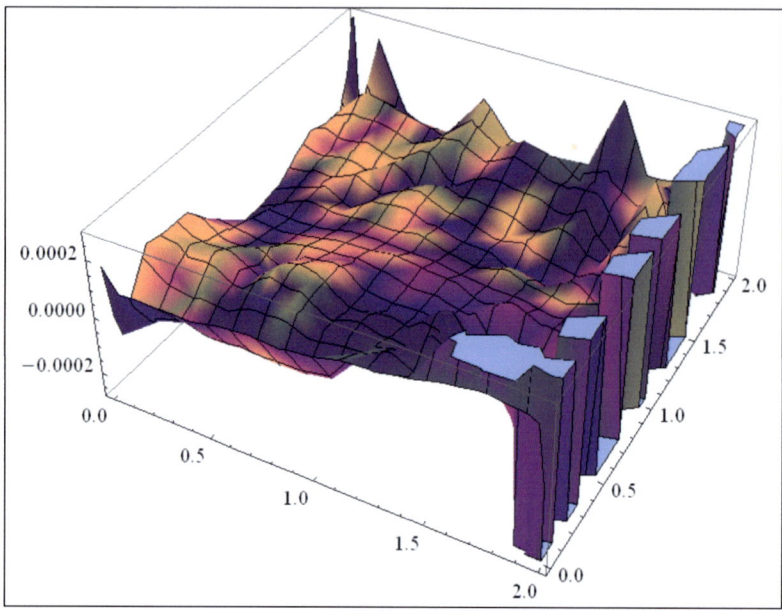

Figure 2

And here both graphs together:

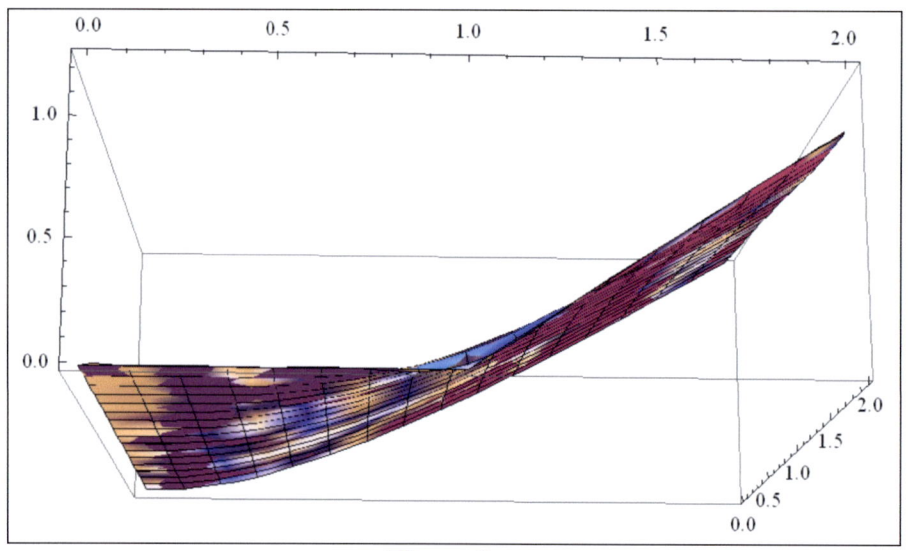

Figure 3

If we estimate only σ, then we get the following results:

j	k_{max}	Q_{min}	Q_2	sse	$\|p - \hat{p}\|$
0	3	0.00429676	0.00950437	0.00792428	0.000308146
0	4	0.00122122	0.00348848	0.000727726	0.000171011
0	5	0.00092719	0.00391767	0.000895067	0.0000219414
0	6	0.000881867	0.00381313	0.00116689	0.000129982
1	3	0.00545202	0.0139091	0.00701056	0.00046983
1	4	0.00102757	0.00378992	0.000790042	0.000224114
1	5	0.000450081	0.00505107	0.00276614	0.0000370898
1	6	0.000333051	0.366122	0.0516826	9.64761×10^{-6}
2	3	83.6402	113.45	2020.25	0.79856
2	4	2.52266	6.5229	28.0959	0.799667
2	5	0.00107984	0.0232849	0.00330719	0.000144208
2	6	0.000172318	0.665764	0.408544	7.18154×10^{-6}

Here is the graph of $v - g$ for $j = 0$ and $k_{max} = 6$ (with the small Q_2 and *sse*):

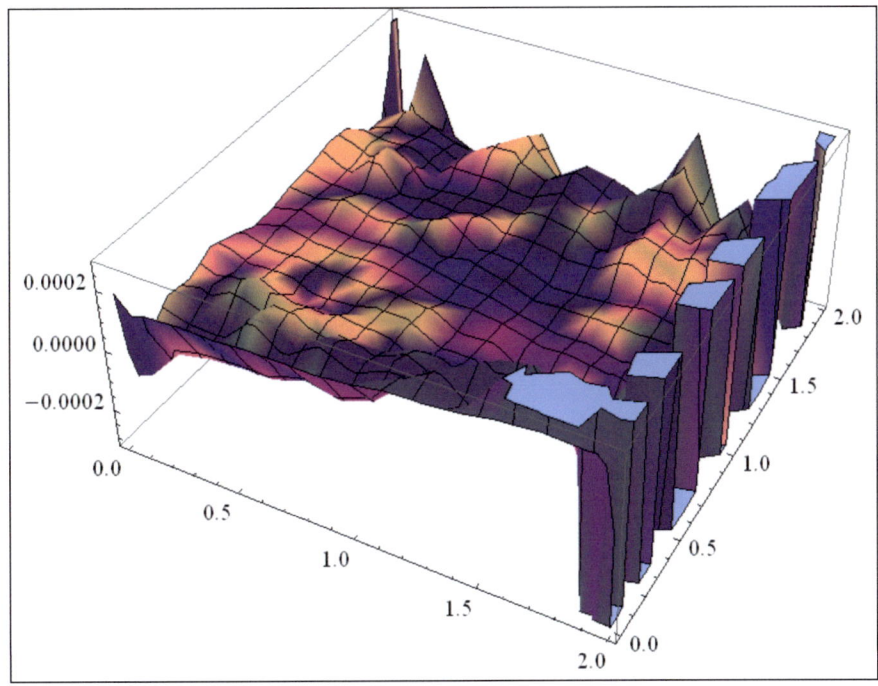

Figure 4

And here both graphs together:

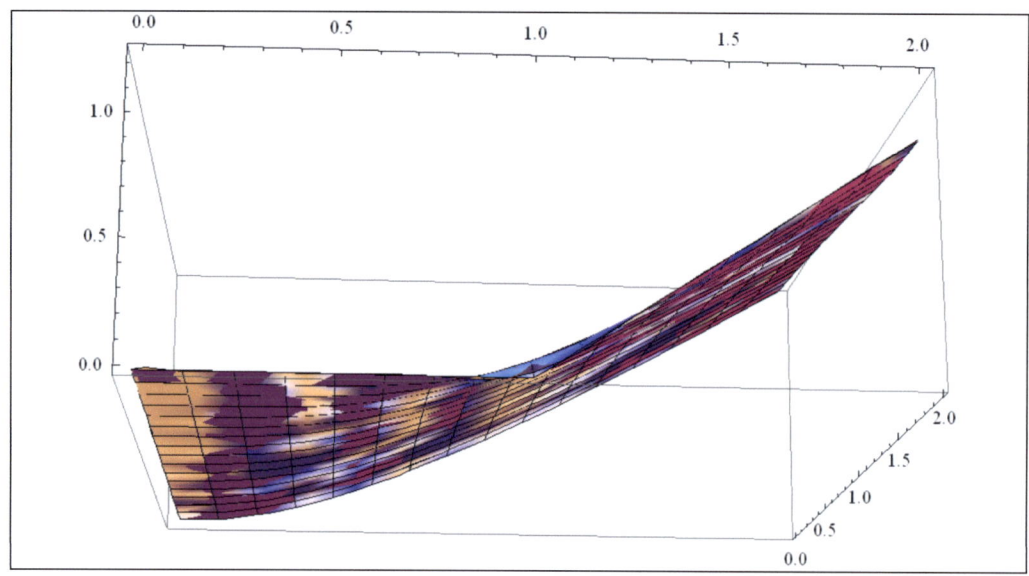

Figure 5

The Algorithm for PDEs

In [10] the algorithm to adjust the parameter of the collocation or minimal residual method was described for ODEs. We used a similar algorithm for the parameter identification in the Balck Scholes PDE. The problem with a PDE is generally, that if we double $n_o - n_u + 1$ and $m_o - m_u + 1$ in for the summation area of the approximation function g, we square the number of basis parameters (the coefficients c_{k_1,k_2}). The same applies to the measurement and collocation points. So we must be careful if we increase the parameters for the summation areas.

If $Q_{min} > \varepsilon_1$ than $j = j + 1$ and $k_{max} = k_{max} + 1$. And in every second minimization step or if $Q_{min} \leq \varepsilon_1$ and $Q_2 > \varepsilon_2$ we halve the step size of the collocation points. The algorithm stops, if $Q_{min} \leq \varepsilon_1$ and $Q_2 \leq \varepsilon_2$ or if a maximal number of iteration (minimization) steps was reached.

At the beginning we set $k_{max} = 4$ and $j = 0$. We set $\varepsilon_1 = 10^{-3}$ and $\varepsilon_2 = 10^{-2}$.

We used the same measurement points like in the chapter before and the same parameters for the simulation of the measurement points. After 2 steps the algorithm stopped ($Q_{min} \leq \varepsilon_1$ and $Q_2 \leq \varepsilon_2$) and we got $\hat{\sigma} = 0.800035$ ($\sigma = 0.8$).

Table of the iteration:

j	$k_{max}^{(0)}$	v	Q_{min}	Q_2	sse
0	4	1	0.00122488	0.00350273	0.0000419011
1	5	1	0.000450072	0.00507484	0.0000372334

sse is the sum of squared errors:

$$sse = \sum_{i=0}^{15}\sum_{j=0}^{15}(v(i/16,j/16)-g(i/16,j/16))^2$$

Graph of g and the exact solution v ($j = 1$, $k_{max} = 5$):

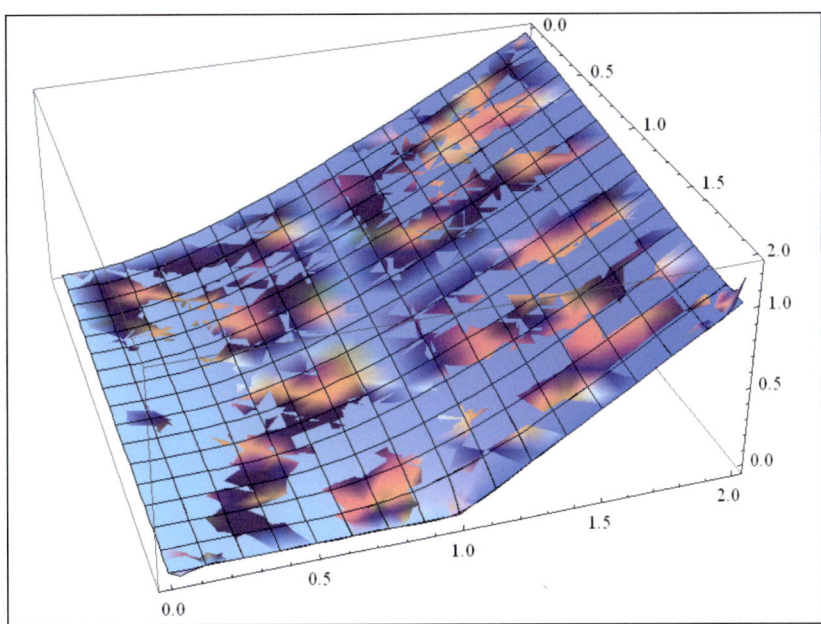

Figure 6

Graph of $v - g$:

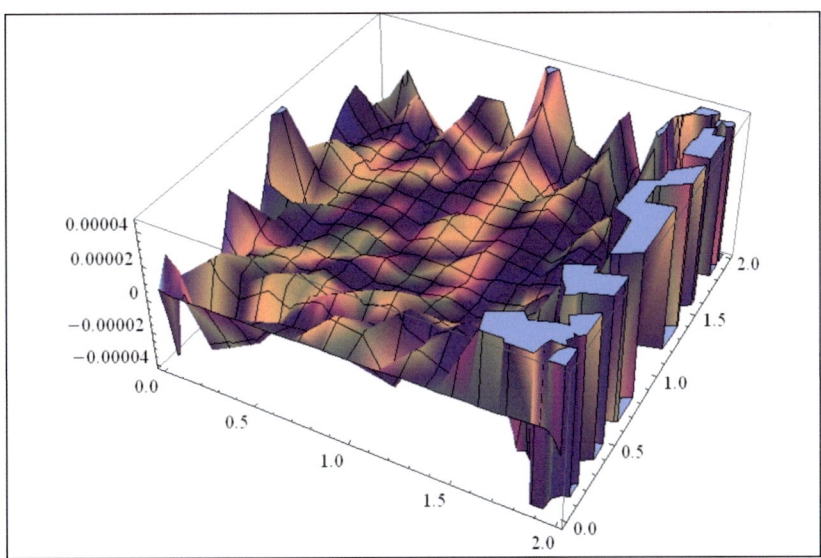

Figure 7

Estimation in two Steps

In [v9] is described a parameter estimation in two steps for an ODE. Here we apply this method on the Black Scholes PDE.

In the first step we estimate c

$$\min_{c} Q_{10}(c)$$

with

$$Q_{10}(c) = \sum_{i}\sum_{l}(\tilde{m}_{i,l} - g(\hat{s}_i, \hat{t}_l))^2 .$$

Then we use the estimation of c from step 1 in the function g.
In the second step we estimate p

$$\min_{c,p} Q_{01}(p)$$

with

$$Q(p) = \sum_{i}\sum_{l} F(g(s_i,t_l), g_s(s_i,t_l), g_t(s_i,t_l), g_{ss}(s_i,t_l), s_i, t_l)^2 .$$

We use the Black Scholes PDE:

$$F(v, v_s, v_t, v_{ss}, s, t) = g_t + 1/2 \cdot \sigma^2 s^2 \cdot g_{ss} + r \cdot s \cdot g_s - r \cdot g = 0$$

Additionally we could use the boundary conditions

$$g(s,T) = \underbrace{1_{[K,\infty)}(s) \cdot (s-K)}_{:=R(s)}$$

in the first step.

Here in the second step we need only a no intercept regression through the points (at the measurement points (1)).

$$(1/2 \cdot s^2 \cdot g_{ss}, -(g_t + r \cdot s \cdot g_s - r \cdot g)),$$

so the slope is $\hat{\sigma}^2$.

We used the following measurement and collocation points:
We simulated measurements $\tilde{m}_{i,l}$ from at the points $(\hat{s}_i, \hat{t}_l) \in [0, 2] \times [0, T]$ with $T = 2$. We set $K = 1$ and for the simulation we use $r = 0.01$ and $\sigma = 0.8$. As measurement points we choose

(1) $\hat{s}_i = i \cdot 1/20$, $i = 0,\ldots,39$, $\hat{t}_i = i \cdot 1/20$ $i = 0,\ldots,39$

and we use the same collocation points

(2) $s_i = i \cdot 1/10$, $i = 0,\ldots,19$, $t_i = i \cdot 1/10$ $i = 0,\ldots,19$.

The unknown parameter vector is only the volatility σ.

For the approximation function we used $j = 1$ and $n_0 = m_0 = 2k_{max}$, $n_u = m_u = -k_{max}$ and with $k_{max} = 6$.

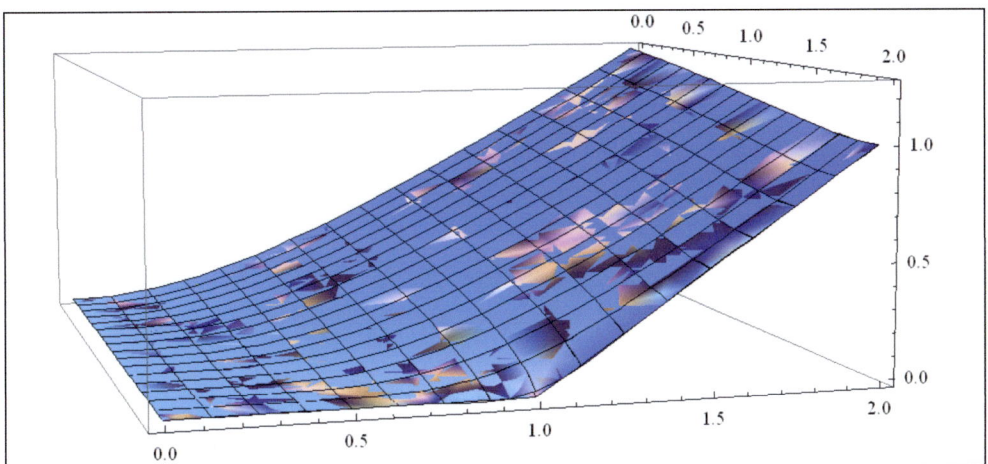

Figure 8

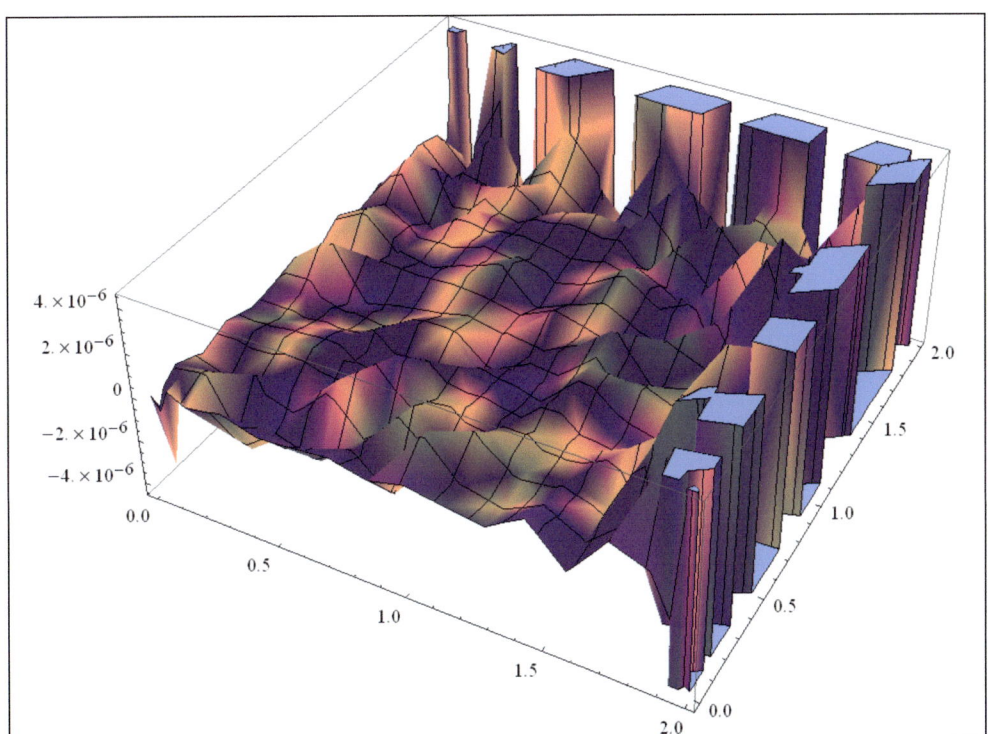

Figure 9

We got $\hat{\sigma}^2 = 0.800042\ldots$.

Remark:

The exact solution of the Black Scholes PDE is known, so we could apply a "direct" approximation of the volatility by minimization if the sum of

$$\sum_i \sum_l (\tilde{m}_{i,l} - v(\hat{s}_i, \hat{t}_l))^2.$$

The Dupire Formula

Many papers use the formula of Dupire. Here the dual Black Scholes PDE is dissolved after the volatility (see [13]):

$$\hat{\sigma}^2 = \frac{v^*_t(s^*, t^*) + r \cdot s \cdot v_s(s^*, t^*)}{1/2 \cdot s^2 \cdot v_{ss}(s^*, t^*)}$$

Here t^* is a fixed time, s^* a fixed price v^* is the observed value.

References

[1] T. S. Carlson, J. Dockery, J. Lund. *A Sinc-Collocation Method for Initial Value Problems.* Mathematics and Computation, Vol. 66, No. 217 (1997)

[2] S. Kameswaran, L. T. Biegler. *Simultaneous Dynamic Optimization Strategies: Recent Advances and Challenges.* Computers and Chemical Engineering, (2006).

[3] A. Nurmuhammada, M. Muhammada, M. Moria, M. Sugiharab. *Double Exponential Transformation in the Sinc-Collocation Method for a Boundary Value Problem with Fourth-Order Ordinary Differential Equation.* Journal of Computational and Applied Mathematics, (2005).

[4] L. Qian. *On the Regularized Whittaker-Koltel'nikov-Shannon Sampling Theorem.* Proceedings of the Amarican Mathematical Society, Vol. 131, No. 4, (2002)

[5] M. Schuchmann. *Approximation and Collocation with Wavelets. Approximations and Numerical Solving of ODEs, PDEs and IEs*. Osnabrück: DAV, (2012).

[6] M. Schuchmann, M. Rasguljajew; (2013). *Error Estimation of an Approximation in a Wavelet Collocation Method.* Journal of Applied Computer Science & Mathematics, No. 14 (7) / 2013, Suceava. (http://jacs.usv.ro/index.php?pag=showcontent&issue=14&year=2013).

[7] M. Schuchmann, M. Rasguljajew; (2013). *Error Estimation and Assessment of an Approximation in a Wavelet Collocation Method.* American Journal of Computational Mathematics, Vol.3, No.2, June 2013.

[8] M. Schuchmann, M. Rasguljajew; (2013). *Parameter Identification with a Wavelet Collocation Method in a Partial Differential Equation.* Journal of Approximation Theory and Applied Mathematics, Vol. 1.

[9] M. Schuchmann, M. Rasguljajew; (2013). *An Approach for a Parameter Estimation with a Wavelet Collocation Method.* Journal of Approximation Theory and Applied Mathematics, Vol. 1.

[10] M. Schuchmann, M. Rasguljajew; (2013). *Determination of Optimal Parameters in a Wavelet Collocation Method.* International Journal of Emerging Technology and Advanced Engineering, Vol. 3, Issue 5, May 2013. (http://www.ijetae.com/files/Volume3Issue5/IJETAE_0513_01.pdf)

[11] M. Schuchmann, M. Rasguljajew; (2013). *Comparing Approximations of a Wavelet Collocation Method of Various Wavelets*. Journal of Approximation Theory and Applied Mathematics, Vol. 2.

[12] M. Schuchmann, M. Rasguljajew; (2013). *Implementation and Testing an Algorithm for a Wavelet Collocation Method in Mathematica.* International Journal of Emerging Technology and Advanced Engineering, Vol. 3, Issue 6, June 2013. (http://www.ijetae.com/files/Volume3Issue6/IJETAE_0613_01.pdf)

[13] Rückert, N. Anderssen, R. S. Hofmann, B. (2012). "Stable Parameter Identification Evaluation of Volatility", Preprint 2012-3, *Preprintreihe der Fakultät f ur Mathematik ISSN 1614-8835*

[14] M. Unser. *Vanishing Moments and the Approximation power of Wavelet Expansions.* Proceedings of the 1996 IEEE International Conference on Image Processing, (1996)

[15] R. Vuduc. *A Wavelet Collocation Method for Solving PDEs.* J. Comp. Phys., (2000).

ADAPTED LINEAR APPROXIMATION FOR LOGARITHMIC KERNEL INTEGRALS

MOSTEFA NADIR

ABSTRACT. In this work we present an approximation for singular integrals with logarithmic kernel on a smooth oriented contour, for this latter we use a small modification of the linear spline functions in order to eliminate the weak singularity. Noting that this approximation is destined to solve numerically the integral equations with weakly singular kernel on a smooth oriented contour.

Introduction

As we know Fredholm integral equations of the second kind appear in many applications among those transport theory, potential theory, elasticity and crack mechanics. All those areas of mathematical physics contain a problems lead to the equation of the form

$$(1) \qquad \varphi(t_0) + \frac{1}{\pi i} \int_\Gamma k(t, t_0)\varphi(t)dt = f(t_0),$$

for a given function $f(t_0)$ and a given singular kernel $k(t, t_0)$ of the type

$$k(t, t_0) = h(t, t_0) \ln(t - t_0),$$

where $h(t, t_0)$ is non singular function and under Γ we designate an oriented smooth contour, the points t and t_0 are on Γ.

Our problem describe a new approximation of singular integral with logarithmic kernel

$$(2) \qquad F(t_0) = \frac{1}{\pi i} \int_\Gamma \varphi(t) \ln(t - t_0)dt, \quad t, t_0 \in \Gamma.$$

Noting that, this integral can be derived when we use the single layer potential for a boundary element method of Laplace equation with the Dirichlet boundary data. for the existence of the principal value of this integral for a given density $\varphi(t)$, we will need more than mere

---
2000 *Mathematics Subject Classification.* Primary 45D05, 45E05, 45L05; Secondary 65R20.

Key words and phrases. Singular integral, interpolation, weakly singular kernel.

continuity. In other words, the density $\varphi(t)$ has to satisfy the Holder condition $H(\mu)[1-2-3]$.

The function $\varphi(t)$ will be said to satisfy a Holder condition on Γ, if for any two points t_1 and t_2 of Γ

$$\mid \varphi(t_2) - \varphi(t_1) \mid \leq A \mid t_2 - t_1 \mid^{\mu}, \quad 0 < \mu \leq 1,$$

where A is a positive constant, called the Holder constant and μ the Holder index.

The Quadrature

We denote by t the parametric complex function $t(s)$ of the curve Γ defined by

$$t(s) = x(s) + iy(s), \quad a \leq s \leq b,$$

where $x(s)$ and $y(s)$ are continuous functions on the finite interval of definition $[a, b]$ and have continuous first derivatives $x'(s)$ and $y'(s)$ never simultaneously null. Let N be an arbitrary natural number, generally we take it large enough and divide the interval $[a, b]$ into N equal subintervals $I_1, I_2, ..., I_N$ by the points

$$s_\sigma = a + \sigma \frac{l}{N}, \quad l = b - a, \quad \sigma = 0, 1, 2, ..., N.$$

Assuming that, for the indices $\sigma, \nu = 0, 1, 2, ..., N - 1$, the points t and t_0 belong respectively to the arcs $\widehat{t_\sigma t_{\sigma+1}}$ and $\widehat{t_\nu t_{\nu+1}}$ where $\widehat{t_\alpha t_{\alpha+1}}$ designates the smallest arc with ends t_α and $t_{\alpha+1}$ $[4-5-6]$.

For an arbitrary number $\sigma = 0, 1, 2, ..., N - 1$, we define the linear spline interpolation polynomial $S_1(\varphi; t, \sigma)$ dependent on φ, t and σ which represents the linear approximation of the function density $\varphi(t)$ on the subinterval $[t_\sigma, t_{\sigma+1}]$ of the curve Γ, given by the following equation

For $t_\sigma \leq t \leq t_{\sigma+1}$,

$$(3) \quad S_1(\varphi; t, \sigma) = \frac{(t_{\sigma+1} - t)}{(t_{\sigma+1} - t_\sigma)} \varphi(t_\sigma) + \frac{(t - t_\sigma)}{(t_{\sigma+1} - t_\sigma)} \varphi(t_{\sigma+1}).$$

This spline function exists and is unique also called a linear interpolating polynomial.

We define for an arbitrary numbers σ and ν such that $0 \leq \sigma, \nu \leq N - 1$, the function $\beta_{\sigma\nu}(\varphi; t, t_0, \sigma, \nu)$ dependents of φ, t and t_0

$$(4) \quad \beta_{\sigma\nu}(\varphi; t, t_0, \sigma, \nu) = \begin{cases} U(\varphi; t, t_0, \sigma) - V(\varphi; t, t_0, \sigma, \nu) & \text{if } t \neq t_0 \\ 0 & \text{if } t = t_0 \end{cases}$$

The function $U(\varphi;t,t_0,\sigma)$ represents a modified linear interpolation of the function density $\varphi(t)$ on the subinterval $[t_\sigma, t_{\sigma+1}]$ of the curve Γ. Indeed, for $t_\sigma \leq t \leq t_{\sigma+1}$ and $t - t_0 \neq 1$, we put

$$U(\varphi;t,t_0,\sigma) = \frac{(t_{\sigma+1}-t)}{(t_{\sigma+1}-t_\sigma)}\varphi(t_{\sigma k})\frac{\ln(t_\sigma - t_0)}{\ln(t-t_0)} + \frac{(t-t_\sigma)}{(t_{\sigma+1}-t_\sigma)}\varphi(t_{\sigma+1})\frac{\ln(t_{\sigma+1}-t_0)}{\ln(t-t_0)},$$

and the function $V(\varphi;t,t_0,\sigma,\nu)$ is given by

$$V(\varphi;t,t_0,\sigma,\nu) = S_1(\varphi;t_0,\nu)\frac{\ln(t_\sigma-t_0)}{\ln(t-t_0)}\frac{(t_{\sigma+1}-t)}{(t_{\sigma+1}-t_\sigma)} + S_1(\varphi;t_0,\nu)\frac{\ln(t_{\sigma+1}-t_0)}{\ln(t-t_0)}\frac{(t-t_\sigma)}{(t_{\sigma+1}-t_\sigma)},$$

where the function φ represents a given function on the curve Γ of the class $H(\mu)$.

Denoting by $\psi_{\sigma\nu}(\varphi;t,t_0,\sigma,\nu)$ the quadratic approximation of the density $\varphi(t)$ at the point $t \in [t_\sigma, t_{\sigma+1}]$, $t_0 \in [t_\nu, t_{\nu+1}]$ and $0 \leq \sigma, \nu \leq N-1$ by

(5) $$\psi_{\sigma\nu}(\varphi;t,t_0,\sigma,\nu) = \varphi(t_0) + \beta_{\sigma\nu}(\varphi;t,t_0,\sigma,\nu),$$

and replacing the density $\varphi(t)$ by expansion (5) in the weakly singular integral (2)

$$F(t_0) = \frac{1}{\pi i}\int_\Gamma \varphi(t)\ln(t-t_0)dt,$$

and obtain the following approximation noting by $S(\varphi;t)$ given as

(6) $$\begin{aligned}S(\varphi;t_0) &= \frac{1}{\pi i}\int_\Gamma \psi_{\sigma\nu}(\varphi;t,t_0,\sigma,\nu)\ln(t-t_0)dt \\ &= \frac{1}{\pi i}\int_\Gamma \beta_{\sigma\nu}(\varphi;t,t_0,\sigma,\nu)\ln(t-t_0)dt.\end{aligned}$$

Main results

Theorem
Let Γ be an oriented smooth contour and let φ be a density function defined on Γ and satisfying the Holder condition $H(\mu)$ then, the

following estimation

$$| F(t_0) - S(\varphi; t_0) | \leq \frac{C \ln N}{N^\mu} \quad N > 1$$

holds, where the constant C depends only of the contour Γ and the Holder's constant.

Proof

For any points $t \in [t_\sigma, t_{\sigma+1}]$ and $t_0 \in [t_\nu, t_{\nu+1}]$ we have

(7)
$$\begin{aligned}\varphi(t) - \psi_{\sigma\nu}(\varphi; t, t_0, \sigma, \nu) &= \varphi(t) - \varphi(t_0) \\ &- \{ \frac{t_{\sigma+1} - t}{t_{\sigma+1} - t_\sigma} \varphi(t_\sigma) \frac{\ln(t_\sigma - t_0)}{\ln(t - t_0)} \\ &+ \frac{t - t_\sigma}{t_{\sigma+1} - t_\sigma} \varphi(t_{\sigma+1}) \frac{\ln(t_{\sigma+1} - t_0)}{\ln(t - t_0)} \\ &- \frac{S_1(\varphi; t_0, \nu) \ln(t_\sigma - t_0)(t_{\sigma+1} - t)}{\ln(t - t_0)(t_{\sigma+1} - t_\sigma)} \\ &- \frac{S_1(\varphi; t_0, \nu) \ln(t_{\sigma+1} - t_0)(t - t_\sigma)}{\ln(t - t_0)(t_{\sigma+1} - t_\sigma)} \}.\end{aligned}$$

Taking into account the expression (7) we get

(8)
$$\int_\Gamma \ln(t - t_0)(\varphi(t) - \psi_{\sigma\nu}(\varphi; t, t_0, \sigma, \nu))dt = \sum_{\sigma=0}^{N-1} \int_{t_\sigma t_{\sigma+1}} (\varphi(t) - \varphi(t_0)) \ln(t - t_0) - \beta_{\sigma\nu}(\varphi; t, t_0, \sigma, \nu) \ln(t - t_0) dt,$$

hence

$$\begin{aligned}F(t_0) - S(\varphi; t_0) = \frac{1}{\pi i} \sum_{\sigma=0}^{N-1} \int_{t_\sigma t_{\sigma+1}} &(\varphi(t) - \varphi(t_0)) \ln(t - t_0) \\ -\{ \frac{t_{\sigma+1} - t}{t_{\sigma+1} - t_\sigma} &\varphi(t_\sigma) \frac{\ln(t_\sigma - t_0)}{\ln(t - t_0)} \\ + \frac{t - t_\sigma}{t_{\sigma+1} - t_\sigma} &\varphi(t_{\sigma+1}) \frac{\ln(t_{\sigma+1} - t_0)}{\ln(t - t_0)} \\ - \frac{S_1(\varphi; t_0, \nu) \ln(t_\sigma - t_0)(t_{\sigma+1} - t)}{\ln(t - t_0)(t_{\sigma+1} - t_\sigma)} & \\ - \frac{S_1(\varphi; t_0, \nu) \ln(t_{\sigma+1} - t_0)(t - t_\sigma)}{\ln(t - t_0)(t_{\sigma+1} - t_\sigma)} &\} \ln(t - t_0) dt.\end{aligned}$$

Seeing that, the equalities $t_\sigma - t_0 = 0$ and $t_{\sigma+1} - t_0$ are possible only when $\sigma = \nu - 1, \nu + 1$ and ν. For the two first cases the integral (8) exists when t_σ tends to t_0 or $t_{\sigma+1}$ tends to t_0; the other case, if $\sigma = \nu$ we can easily seeing that, the function $\beta_{\sigma\sigma}(\varphi; t, t_0, \sigma, \sigma)$ contains $(t_\sigma - t_0)$ and $(t_{\sigma+1} - t_0)$ as factor, for the points $t, t_0 \in [t_\sigma, t_{\sigma+1}]$ we

write
$$\beta_{\sigma\sigma}(\varphi;t,t_0,\sigma,\sigma) = U(\varphi;t,t_0,\sigma) - V(\varphi;t,t_0,\sigma,\sigma),$$

hence

(9)
$$\beta_{\sigma\sigma}(\varphi;t,t_0,\sigma,\sigma) = \frac{(t_{\sigma+1}-t)\ln(t_\sigma - t_0)}{(t_{\sigma+1}-t_\sigma)\ln(t-t_0)}(\varphi(t_\sigma) - S_1(\varphi;t_0,\sigma))$$
$$+ \frac{(t-t_\sigma)\ln(t_{\sigma+1} - t_0)}{(t_{\sigma+1}-t_\sigma)\ln(t-t_0)}(\varphi(t_{\sigma+1}) - S_1(\varphi;t_0,\sigma)).$$

In other words, we write

$$\beta_{\sigma\sigma}(\varphi;t,t_0,\sigma,\sigma) = \frac{1}{\ln(t-t_0)} Q(\varphi;t,t_0,\sigma,\sigma),$$

where the expression $Q(\varphi;t,t_0,\sigma,\sigma)$ is given by

$$Q(\varphi;t,t_0,\sigma,\sigma) = \frac{(t_{\sigma+1}-t)\ln(t_\sigma - t_0)}{(t_{\sigma+1}-t_\sigma)}(\varphi(t_\sigma) - S_1(\varphi;t_0,\sigma))$$
$$+ \frac{(t-t_\sigma)\ln(t_{\sigma+1} - t_0)}{(t_{\sigma+1}-t_\sigma)}(\varphi(t_{\sigma+1}) - S_1(\varphi;t_0,\sigma)),$$

with

$$\varphi(t_\sigma) - S_1(\varphi;t_0,\sigma) = \frac{(t_0 - t_\sigma)}{(t_{\sigma+1}-t_\sigma)}(\varphi(t_\sigma) - \varphi(t_{\sigma+1})),$$

and

$$\varphi(t_{\sigma+1}) - S_1(\varphi;t_0,\sigma). = \frac{(t_{\sigma+1} - t_0)}{(t_{\sigma+1}-t_\sigma)}(\varphi(t_{\sigma+1}) - \varphi(t_\sigma))$$

Passing now to the estimation of the expression (8).

For $t \in [t_\sigma, t_{\sigma+1}]$ and $t_0 \in [t_\nu, t_{\nu+1}]$ with the conditions $\sigma \neq \nu-1, \nu+1$ and ν we have

$$\left| \frac{1}{\pi i} \sum_{\sigma=0}^{N-1} \int_{t_\sigma t_{\sigma+1}} (\varphi(t) - \varphi(t_0))\ln(t-t_0) \right.$$
$$-\{ \frac{t_{\sigma+1}-t}{t_{\sigma+1}-t_\sigma}\varphi(t_\sigma)\frac{\ln(t_\sigma - t_0)}{\ln(t-t_0)}$$
$$+\frac{t-t_\sigma}{t_{\sigma+1}-t_\sigma}\varphi(t_{\sigma+1})\frac{\ln(t_{\sigma+1} - t_0)}{\ln(t-t_0)}$$
$$-\frac{S_1(\varphi;t_0,\nu)\ln(t_\sigma - t_0)(t_{\sigma+1}-t)}{\ln(t-t_0)(t_{\sigma+1}-t_\sigma)}$$
$$\left. -\frac{S_1(\varphi;t_0,\nu)\ln(t_{\sigma+1} - t_0)(t-t_\sigma)}{\ln(t-t_0)(t_{\sigma+1}-t_\sigma)} \} \ln(t-t_0)dt \right| = O(\frac{\ln N}{N^\mu}).$$

Indeed, it is clear that

$$\max_{t_0 \in \widehat{t_\nu t_{\nu+1}}} \left| \frac{1}{\pi i} \sum_{\sigma=0}^{N-1} \int_{t_\sigma}^{t_{\sigma+1}} (\varphi(t) - \varphi(t_0)) \ln(t - t_0) dt \right| = O\left(\frac{\ln N}{N^\mu}\right)$$

and also we estimate the expression

$$\left| \frac{1}{\pi i} \sum_{\sigma=0}^{N-1} \int_{t_\sigma t_{\sigma+1}} -\left\{ \frac{t_{\sigma+1} - t}{t_{\sigma+1} - t_\sigma} \varphi(t_\sigma) \frac{\ln(t_\sigma - t_0)}{\ln(t - t_0)} \right. \right.$$
$$+ \frac{t - t_\sigma}{t_{\sigma+1} - t_\sigma} \varphi(t_{\sigma+1}) \frac{\ln(t_{\sigma+1} - t_0)}{\ln(t - t_0)}$$
$$- \frac{S_1(\varphi; t_0, \nu) \ln(t_\sigma - t_0)(t_{\sigma+1} - t)}{\ln(t - t_0)(t_{\sigma+1} - t_\sigma)}$$
$$\left. \left. - \frac{S_1(\varphi; t_0, \nu) \ln(t_{\sigma+1} - t_0)(t - t_\sigma)}{\ln(t - t_0)(t_{\sigma+1} - t_\sigma)} \right\} \ln(t - t_0) dt, \right|$$

$$\simeq \left| \frac{1}{\pi i} \sum_{\sigma=0}^{N-1} \int_{t_\sigma t_{\sigma+1}} \frac{\varphi(t_\nu) - \varphi(t_\sigma)}{t_\nu - t_\sigma} + \frac{\varphi(t_{\nu+1}) - \varphi(t_{\sigma+1})}{t_{\nu+1} - t_{\sigma+1}} dt \right| = O\left(\frac{\ln N}{N^\mu}\right).$$

Naturally, the estimation given above is obtained by using the density φ, as an element of the Holder space $H(\mu)[2]$, and the following natural estimation

$$\left| \frac{\ln(t_\sigma - t_0)}{\ln(t - t_0)} \right| \simeq \left| \frac{\ln(t_{\sigma+1} - t_0)}{\ln(t - t_0)} \right| = O(1).$$

Further, for the cases where $\sigma = \nu - 1, \nu + 1$ and ν, using the relation (9) and the smoothness of Γ, we obtain

$$\left| \int_{t_\nu t_{\nu+1}} (\varphi(t) - \varphi(t_0)) \ln(t - t_0) dt \right| \leq A \int_{s_\nu}^{s_{\nu+1}} |s - s_0|^\mu |\ln(s - s_0)| ds$$
$$= O\left(\frac{\ln N}{N^{(\mu+1)}}\right)$$

Numerical experiments

Using our approximation, we apply the algorithms to singular integrals and we present results concerning the accuracy of the calculations, in this numerical experiments each table I represents the exact weakly singular integral and $\widetilde{I}$ corresponds to the approximate calculation produced by our approximation at points values interpolation.

Example 1

Consider the weakly singular integral,

$$I = F(t_0) = \int_\Gamma \varphi(t) \ln(t - t_0) dt,$$

where the curve Γ designate the unit circle and the function density φ is given by the following expression

$$\varphi(t) = -\frac{1}{t^2}.$$

N	$\| I - \tilde{I} \|_1$	$\| I - \tilde{I} \|_2$	$\| I - \tilde{I} \|_\infty$
20	5.7174414E-02	3.5285469E-02	2.9367447E-02
30	5.5649281E-03	2.9125938E-03	1.8216968E-03
40	1.4109910E-03	7.3831965E-04	5.1295757E-04

Example 2

Consider the weakly singular integral,

$$I = F(t_0) = \int_\Gamma \varphi(t) \ln(t - t_0) dt,$$

where the curve Γ designate the unit circle and the function density φ is given by the following expression

$$\varphi(t) = \frac{2}{t^3}.$$

N	$\| I - \tilde{I} \|_1$	$\| I - \tilde{I} \|_2$	$\| I - \tilde{I} \|_\infty$
20	1.5601690E-01	9.2777811E-02	6.5414310E-02
30	1.5781343E-02	9.7699165E-03	6.8665147E-03
40	2.1157265E-03	1.1530236E-03	7.7348948E-04

Note Many examples confirm the efficiency of this approximation.

BIBLIOGRAPHY

[1] **J. ANTIDZE**, On the approximate solution of singular integral equations, Seminar of Institute of Applied Mathematics, 1975, Tbilissi.

[2] **N. I. MUSKHELISHVILI**, Singular integral equations, Naukah Moscow, 1968, English transl, of 1sted Noordho, 1953; reprint,1972.

[3] **M. NADIR, J. ANTIDZE,** On the numerical solution of singular integral equations using Sanikidze's approximation, Comp Math in Sc Tech. 10(1), 83-89 (2004).

[4] **M. NADIR**, Opérateurs intégraux et bases d'ondelettes, Far East J.Sci. 6(6)(1998), 977-995.

[5] **M. NADIR**, Adapted Quadratic Approximation for Singular Integrals, in Journal of mathematical Inequalities 4, (3) pp 423-430 (2010).

[6] **M. NADIR**, Adapted Linear Approximation for Singular Integrals, in Mathematical sciences 6, (36) 2012.

Address. Prof. Dr. Mostefa NADIR
Laboratory of Pure and Applied Mathematics
and
Laboratory of Signals Analysis and Systems
University of Msila
28000 ALGERIA
E-mail. mostefanadir@yahoo.fr

Identifying a Superposition with Trigonometric Functions by Applying a MRA with the Shannon Wavelet

M. Schuchmann and M. Rasguljajew from the Darmstadt University of Applied Sciences

Abstract

The multi resolution analysis (MRA) of the wavelet theory defines a sequence of close subspaces $\{V_j\}_{j \in Z}$ with $V_j \subset V_{j+1} \subset L^2(R)$. The trigonometric functions sin and cos are not quadratic integrable on R. However we can express them with bases functions from V_j by using the Shannon wavelet.

Introduction

In this article we use the Shannon wavelet. For the approximation using the space V_j we even can use functions that are not quadratic integrable on R if we only need an approximation on a finite interval I as in practical case. Considering the interval I, we could use the function $1_I f$ instead of f, if f is quadratic integrable on I (with the indicator function 1) and then $1_I f$ is in $L^2(R)$. But that leads often to worse approximations (see [5], [6] and [7]). For trigonometric functions like sin and cos (or e^{ia}) we can calculate directly the bases coefficients and under certain conditions we can detect a superposition with that trigonometric functions like with the Fourier analysis.

With the scaling function ϕ of the MRA we get an orthonormal basis of V_j with $\phi_{j,k}(t) = 2^{j/2}\phi(2^j t - k)$. So we get the orthogonal projection from a $L^2(R)$ function f in V_j with

$$f_j(t) = \sum_k f_k^j \phi_{j,k}(t) \text{ with } f_k^j = \langle f, \phi_{j,k} \rangle = \int_{-\infty}^{\infty} f(t) \cdot \overline{\phi_{j,k}(t)} dt .$$

In the MRA the spaces V_j are closed subspaces of $L^2(R)$. If f is not quadratic integrabel on R we say that we can "identify" f with V_j, instead of f is in V_j, if we can express f with the orthonormal basis of V_j.

Example:

Let be
$$f(t) = e^{-t^2} + 0.05 \cdot sin(8t) .$$

We show the graph of f together with the approximation f_1:

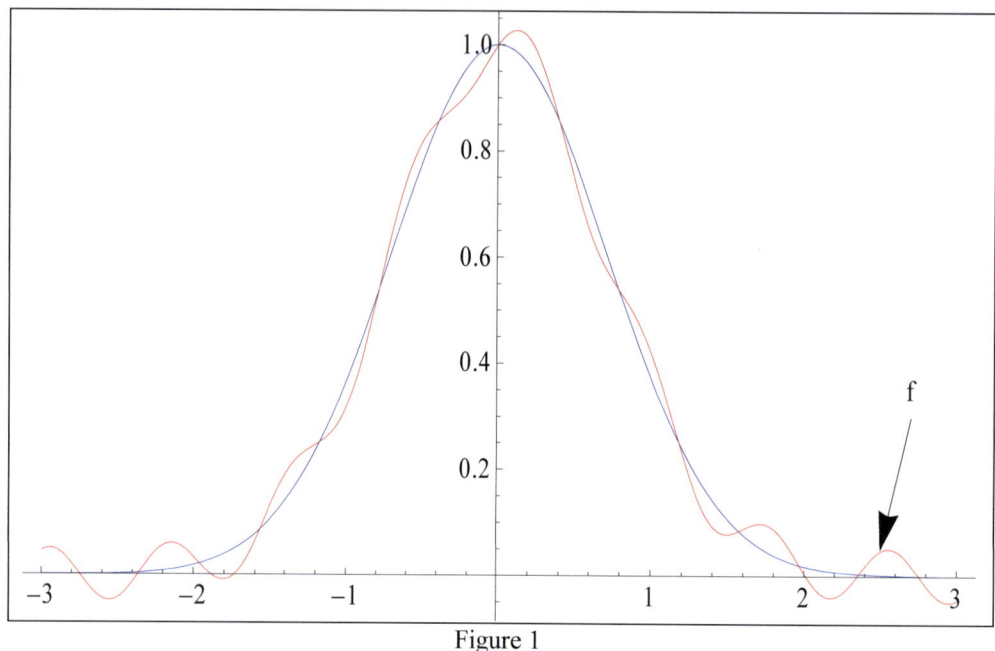

Figure 1

With the function d_1 we can "identify" the superposition term $0.05 \cdot sin(8t)$, what we can see graphically with the graph of d_1 and soon theoretically.

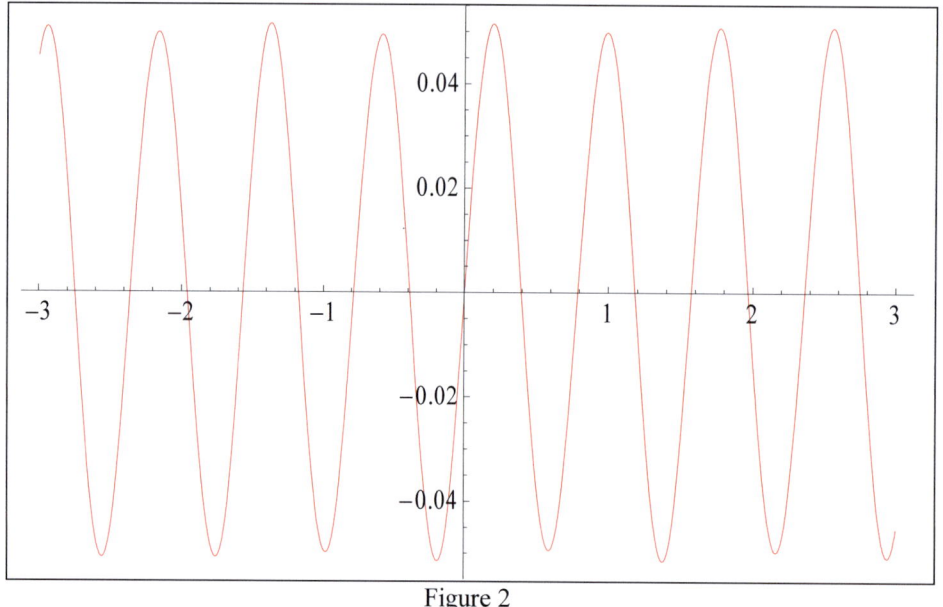

Figure 2

sin(at) and V_j

If we use the Shannon wavelet, f is in V_j if $supp\ F = [-2^j \cdot \pi,\ 2^j \cdot \pi]$ (or if $supp\ F \subseteq [-2^j \cdot \pi,\ 2^j \cdot \pi]$). If f is in detail space W_j then f is in V_{j+1} but not in V_j, because of $V_{j+1} = V_j \oplus W_j$. So if f is quadratic integrable on R then f is in W_j if $supp\ F \subset [-2^{j+1} \cdot \pi,\ -2^j \cdot \pi) \cup (2^j \cdot \pi,\ 2^{j+1} \cdot \pi]$.

In the example above we saw, that we could recognize if a function f is superposed with a sinus function, for example $f(t) = g(t) + c \cdot sin(at)$, when we use the Shannon wavelet in a MRA.

The reason is: The Fourier transform of $h(t) = sin(at)$ is

$$H(\omega) = \frac{1}{\sqrt{2\pi}} \int_{-\infty}^{\infty} h(t) \cdot e^{-i\omega t} dt = i\sqrt{\pi/2} \cdot (\delta(\omega+a) - \delta(\omega-a)),$$

with the Dirac delta distribution δ (using for transformation and back-transformation the factor $1/\sqrt{2\pi}$). So the Fourier transform of $h(t) = sin(at)$ (from now we choose only $a > 0$) is not a function and h is not quadratic integrable on R but we could show that we get for the basis coefficients in Fourier space $<H, \Phi_{j,k}> = 2^{-j/2} h(k/2^j)$ for $a < 2^j \cdot \pi$ and we can show even directly that

$$f_k^j = <h, \phi_{j,k}> = 2^{-j/2} h(k/2^j) \text{ for } a < 2^j \cdot \pi$$

although $h \notin L^2(R)$ (for $a = 2^j \cdot \pi$ all f_k^j vanish). Here we can use the equations

$$\int_{-\infty}^{\infty} e^{-i \cdot a \cdot t} \cdot \phi(t) dt = 1_{[-\pi, \pi]}(a)$$

with the indicator function 1 and

$$sin(a \cdot t) = \frac{1}{2i}(e^{i \cdot a \cdot t} - e^{-i \cdot a \cdot t}).$$

We can show, that the integral above exists and so we would get $f_k^j = <h, \phi_{j,k}> = 2^{-j/2} h(k/2^j)$ for $a < 2^j \cdot \pi$.

Example:
Here are graphs of $h - h_{m,j}$ with

$$h_{m,j}(t) = \sum_{k=-m}^{m} 2^{-j/2} \cdot h(k/2^j) \cdot \phi_{j,k}(t)$$

and $h(t) = sin(at)$ ($j = 1$, $a = 4$, left for $m = 40$ and right for $m = 80$):

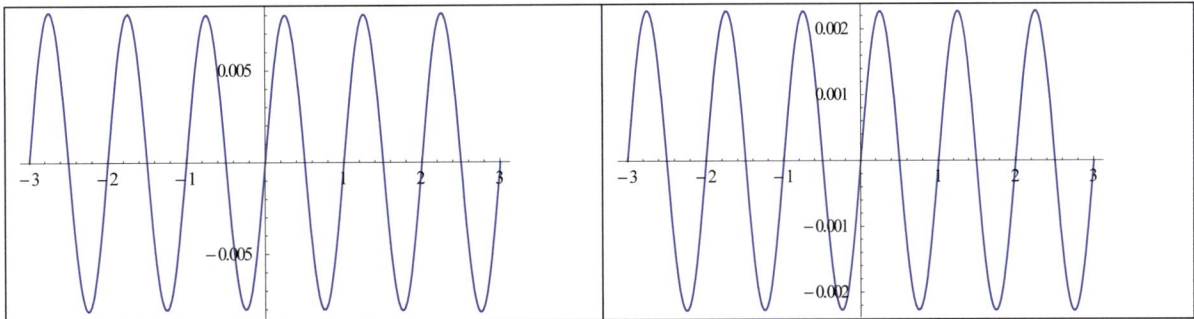

Figure 3

If we would apply the Shannon theorem on h then the condition "$\in L^1(R) \cap L^2(R)$" of the theorem is not met but we can calculate the coefficients of that sinc-series c_k and we would get $c_k = f_k^j$, too, if we set $\Omega = 2^j \cdot \pi$ (for the meaning of Ω see remark at the end of the article).

The angular frequency a must be less than $2^j \cdot \pi$ to identify h with V_j. So we could identify a superposition term $sin(at)$ (or $cos(at)$) with the detail space W_j if $2^j \cdot \pi < a < 2^{j+1} \cdot \pi$. In the first example a was 8, so we could identify the sinus term with $d_1 \in W_1$ because of $2 \cdot \pi < 8 < 4 \cdot \pi$. For the case that $a = 2^j \cdot \pi$: A superposition with $sin(at)$ could not be identified with V_j but with V_{j+1} so $sin(2^j \cdot \pi)$ could be identified with W_j.

Here are graphs (left h and $h_{m,j}$ and right $h - h_{m,j}$) for $m = 40$, $j = 0$ and $a = 1, 2, 3, 4$. We see that $sin(4t)$ could not be identified with V_0.

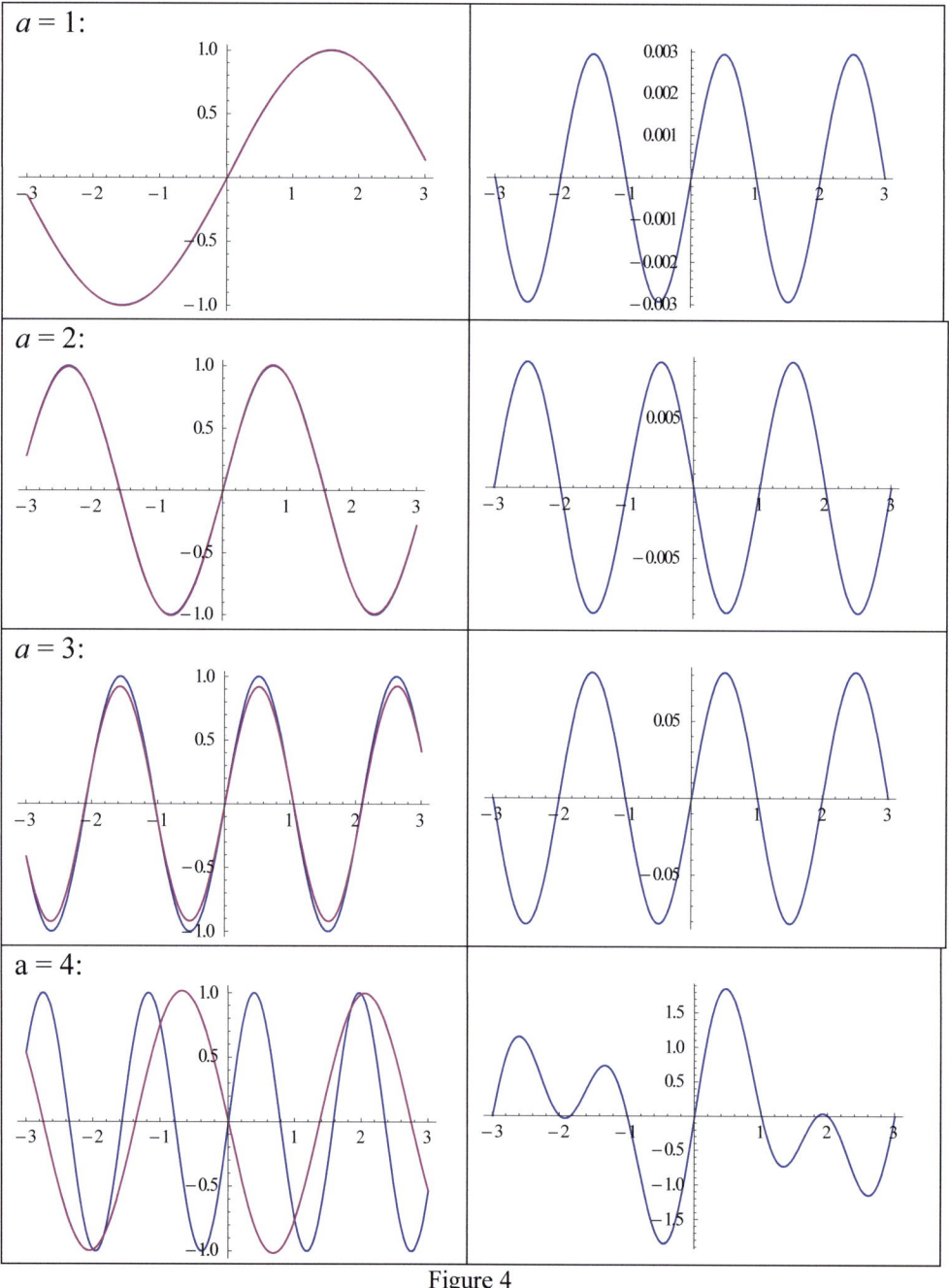

Figure 4

If a is bigger then we need a bigger m, that's what we see with the graph of $sin(3t)$. When we choose $m = 100$ then we get the following graphs:

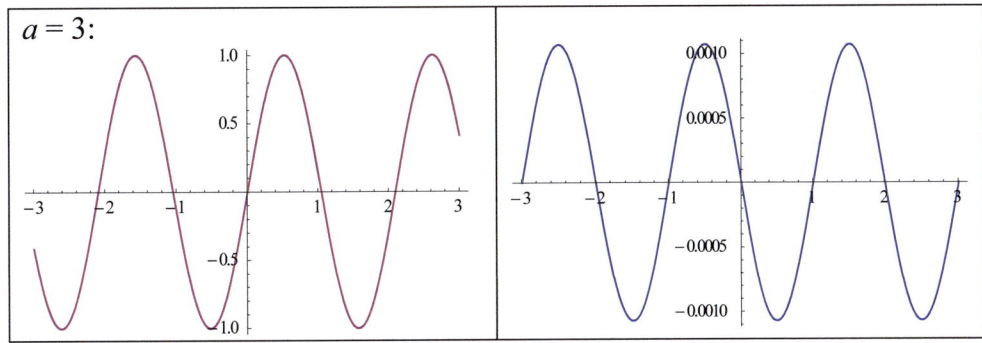

Figure 5

Here are the graphs (left h and $h_{m,j}$ and right $h - h_{m,j}$) for $m = 40$, $j = 2$ and $a = 7, 8, \ldots, 13$. We see that $sin(13t)$ could not be identified with V_2.

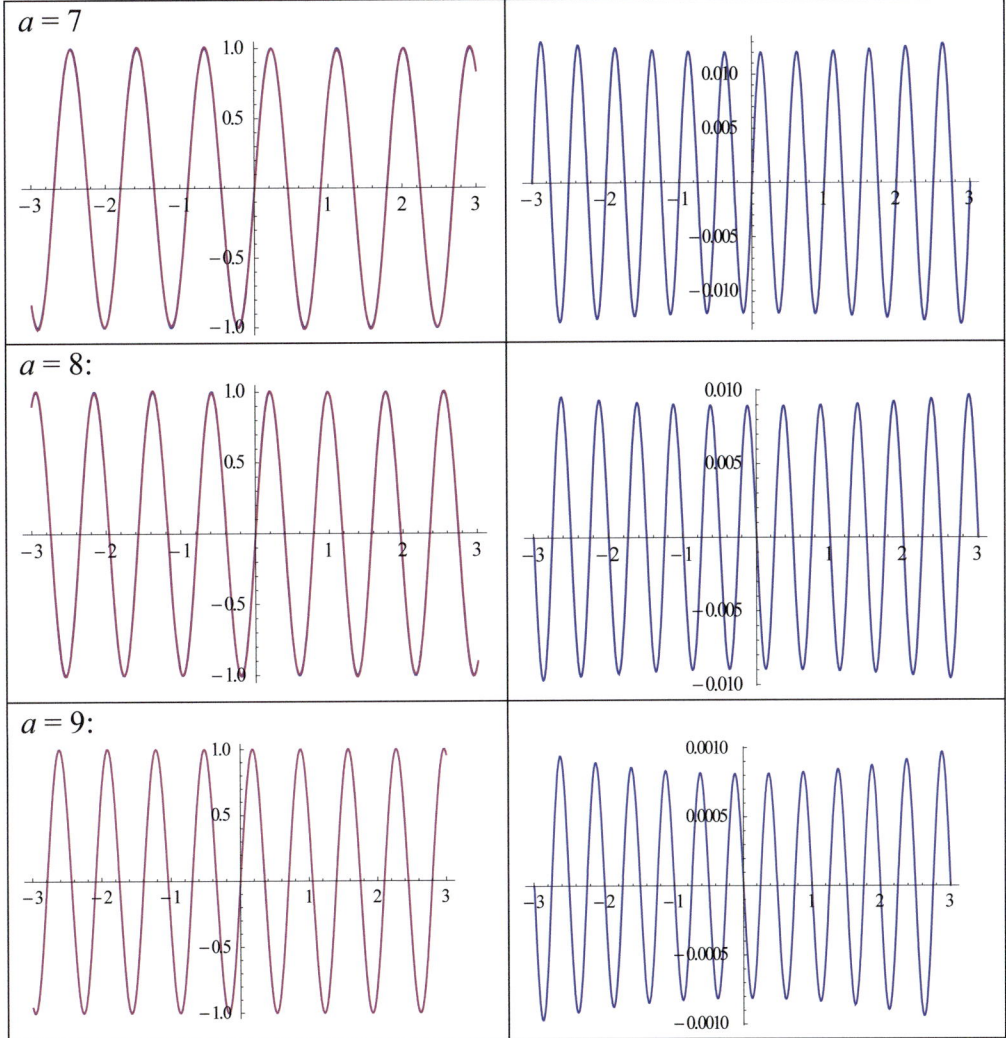

Figure 6a

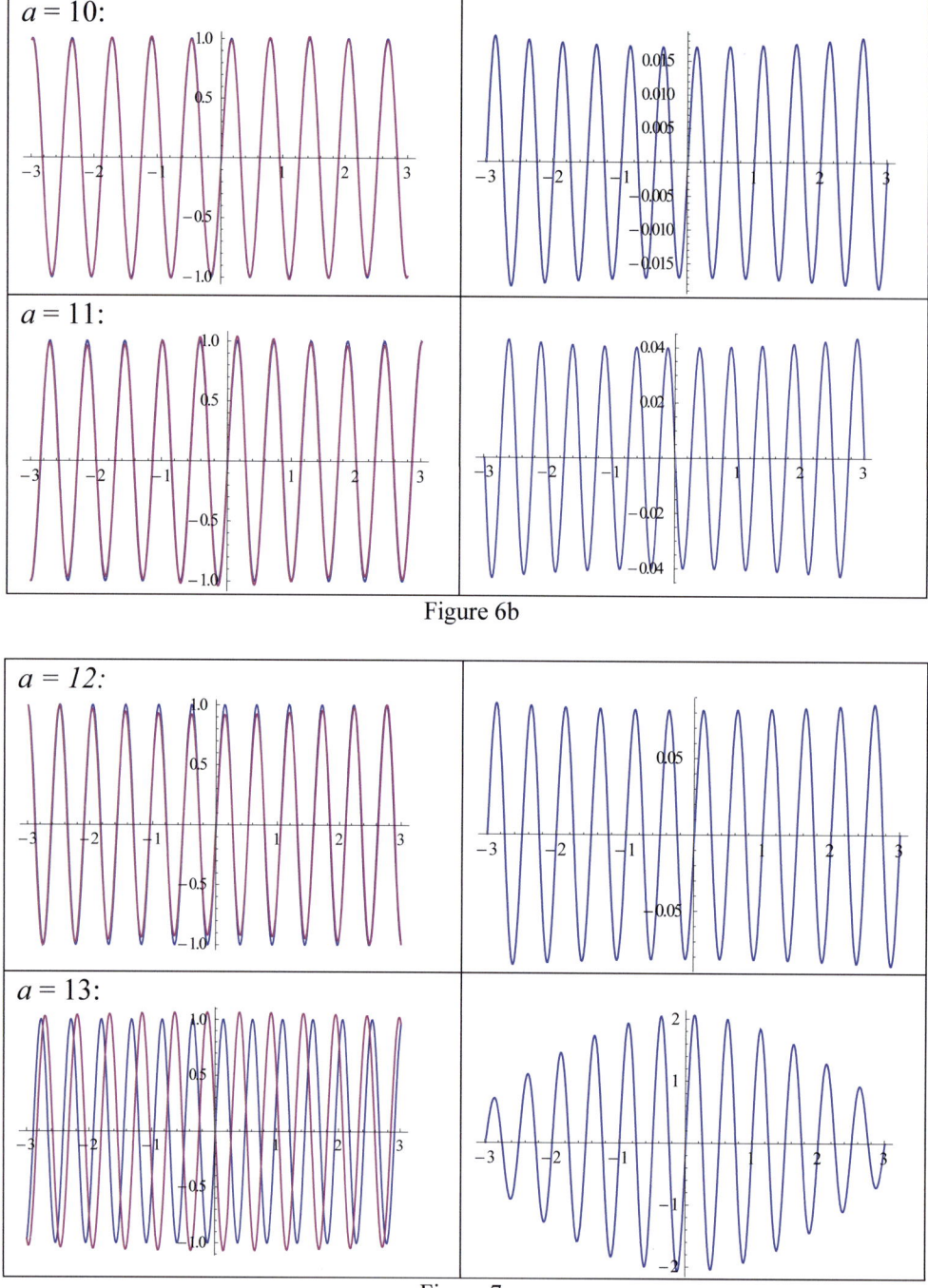

Figure 6b

Figure 7

If we use a function of type $f(t) = g(t) + c \cdot sin(at)$ then it could be possible that we cannot identify the sinus term good in W_j also $2^j \cdot \pi \leq a < 2^{j+1} \cdot \pi$ when the orthogonal projection of g in W_j has a big amount (or when the length I is too small).

Example:

For example if $f(t) = e^{-t^2}$ (which is in $\mathcal{L}^2(R)$) then the graph of d_0 is:

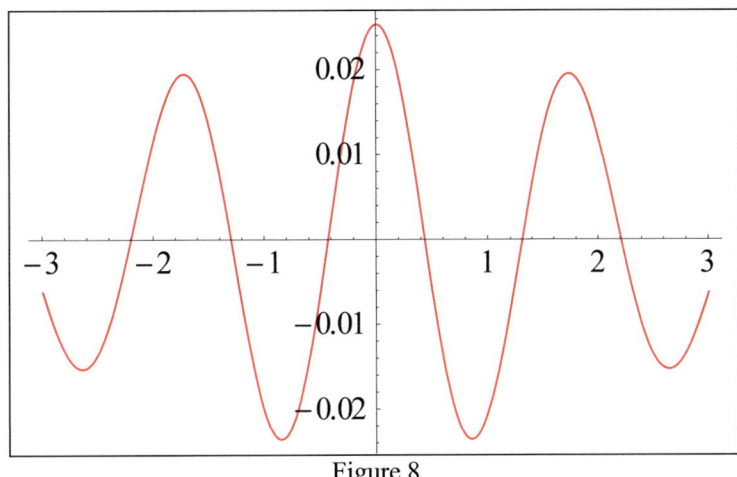

Figure 8

So the orthogonal projection d_0 of f in W_0 is not very small. f is not band-limited but the Fourier transform of f is

$$F(\omega) = \frac{1}{\sqrt{2}} e^{-\omega^2/4}$$

and for example $F(4\pi) \approx 5.06 \cdot 10^{-18}$. So with growing ω the function values $F(\omega)$ becomes "fast" nearly zero as well the detail functions d_j with growing j. That's what we see when we consider the approximation error in Fourier space with the difference of f and f_j (as the orthogonal projection of f in V_j). Here we could calculate the $L^2(I)$ norm $\|f - f_j\|_{L^2(I)}$ with

$$f(t) - f_j(t) = \frac{1}{\sqrt{2\pi}} \int_{-\infty}^{\infty} F(\omega) e^{i\omega t} d\omega - \frac{1}{\sqrt{2\pi}} \int_{-2^j\pi}^{2^j\pi} F(\omega) e^{i\omega t} d\omega$$

$$= \frac{1}{\sqrt{2\pi}} \int_{-\infty}^{-2^j\pi} F(\omega) e^{i\omega t} d\omega + \frac{1}{\sqrt{2\pi}} \int_{2^j\pi}^{\infty} F(\omega) e^{i\omega t} d\omega$$

if we consider the Interval I. For $I = R$ (and on R quadratic integrabel f) we get with the equation from Parseval:

$$\|f - f_j\|_{L^2} = \sqrt{\int_{-\infty}^{-2^j\pi} |F(\omega)|^2 d\omega + \int_{2^j\pi}^{\infty} |F(\omega)|^2 d\omega}$$

So we see that it is important for a good approximation with small j how "fast" $|F(\omega)|$ becomes small with increasing $|\omega|$. If the function f is continuous we could also use the maximum norm.

Analogous we get for $I = R$:

$$\|d_j\|_{L^2} = \sqrt{\int_{-2^{j+1}\pi}^{-2^j\pi} |F(\omega)|^2 d\omega + \int_{2^j\pi}^{2^{j+1}\pi} |F(\omega)|^2 d\omega}$$

Example:

Now we consider the function $f(t) = e^{-t^2} + 0.06\,sin(4t) + 0.02\,sin(10t)$, with the Graph:

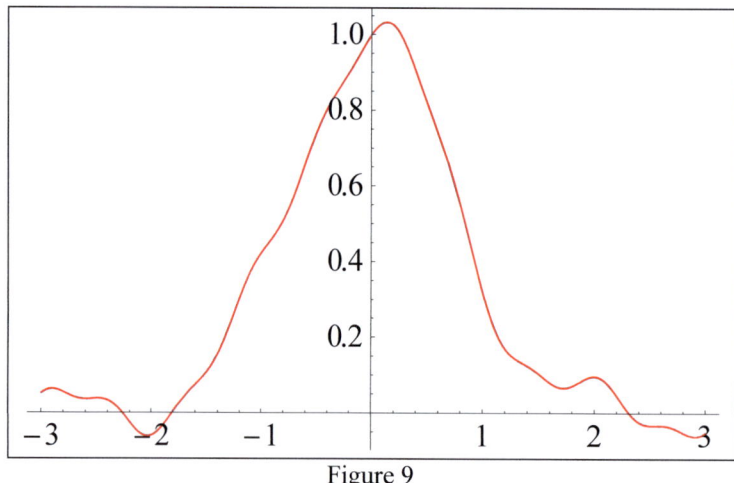
Figure 9

For the following numerical integrations in order to calculate d_j and f_j we used the interval $I = [-30, 30]$. We see no differences between the graph of d_1 and $0.02\,sin(10t)$:

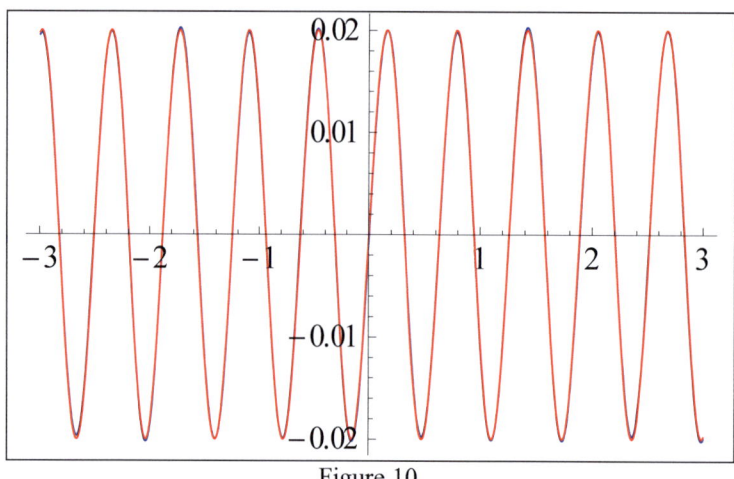
Figure 10

But between the graph of d_0 and $0.06sin(4t)$ (which is dashed) we see a difference:

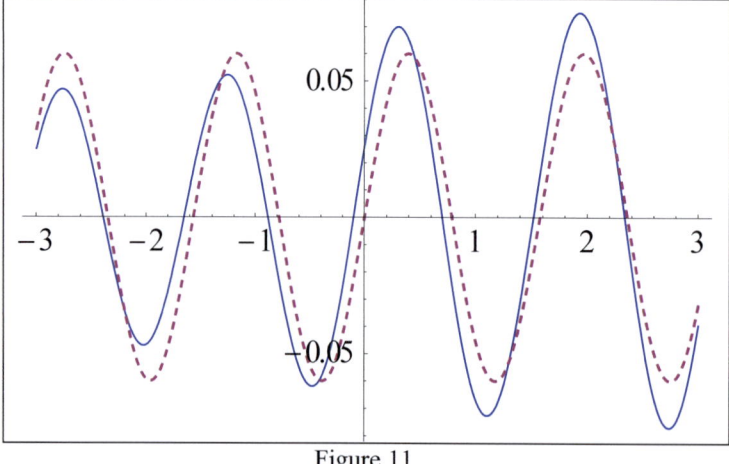
Figure 11

In d_1 the part of the orthogonal projection of e^{-t^2} does not have a big amount, but in d_0.

Here is the graph of f_1 and f (f is dashed):

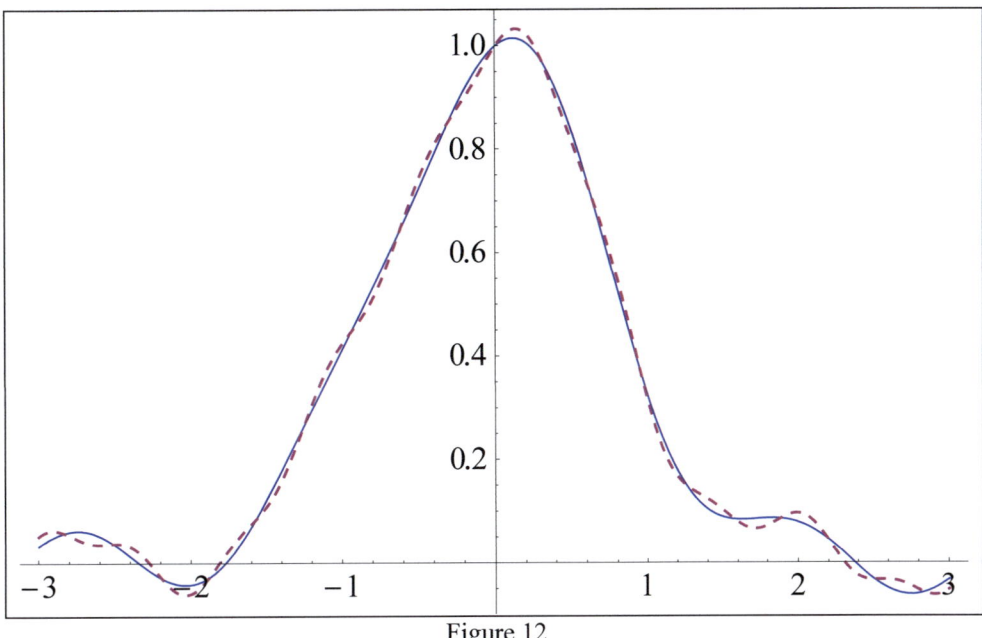

Figure 12

Here is the graph of f_0 and f (f is dashed):

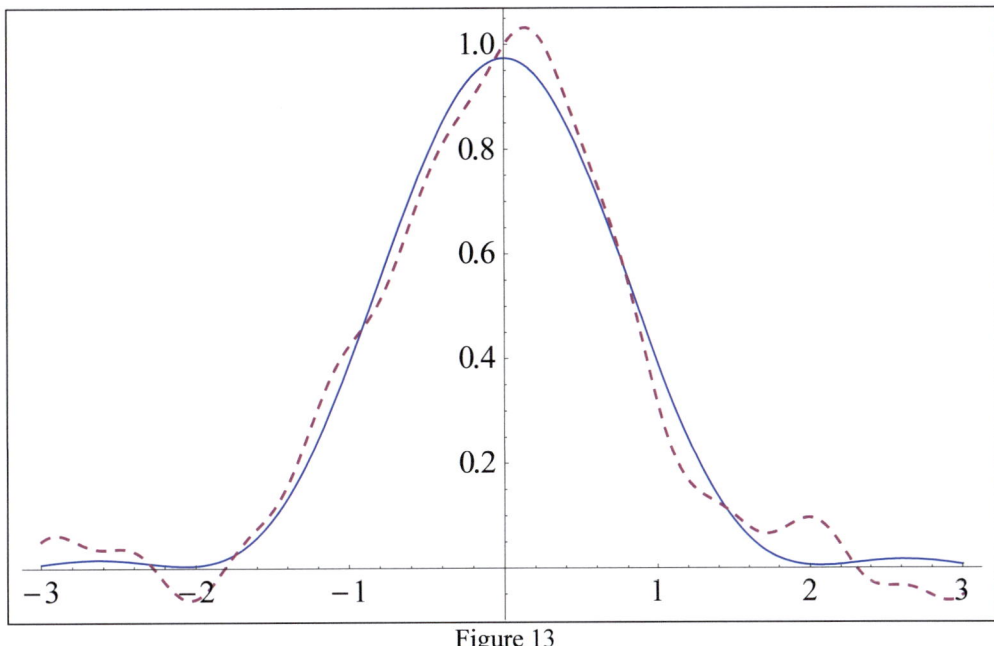

Figure 13

Finally here is the graph of the Fourier transform of $1_I(t) \cdot 0.06 \sin(4t)$ divided by i (for $I = [-30, 30]$) which is concentrated at the points $\omega = \pm 4$ (what is seen even better the bigger the interval I is):

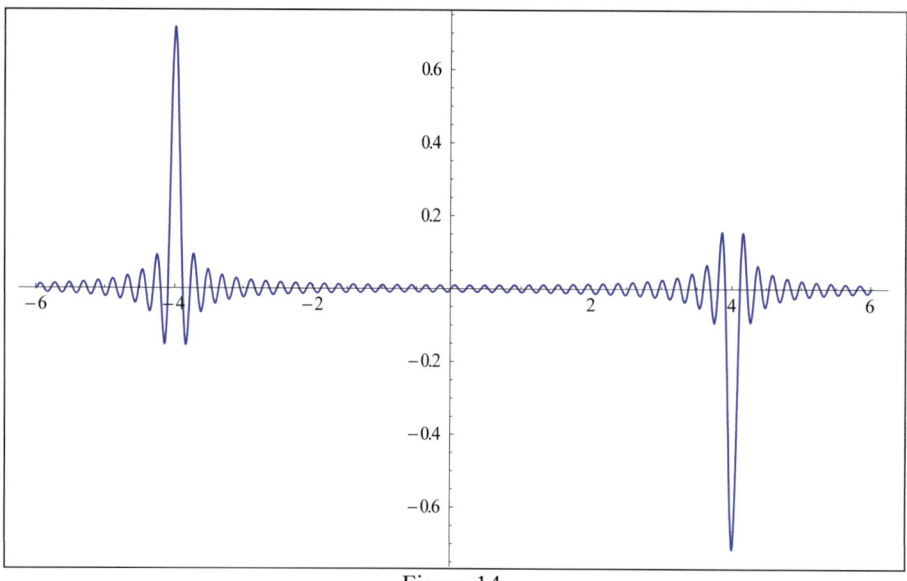

Figure 14

Remark:

We can also show with a different path (what we know from above), that for example $sin(a\cdot\pi\cdot t)$ can be expressed through the bases coefficients of V_j. Here we use the often used notation of the Shannon theorem. In V_j we have $\Omega = 2^j \cdot \pi$.

If the Fourier transform F of f has compact support ($supp\ F \subseteq [-\Omega, \Omega]$) and $f \in \mathcal{L}(R) \cap \mathcal{L}^2(R)$ than

$f(t) = f_s(t)$ (for almost all real t) with

$$f_s(t) = \sum_{k \in Z} f\left(\frac{k \cdot \pi}{\Omega}\right) \cdot \frac{sin(\Omega \cdot t - k \cdot \pi)}{\Omega \cdot t - k \cdot \pi} = \sum_{k \in Z} f\left(\frac{k \cdot \pi}{\Omega}\right) \cdot \frac{sin(\Omega \cdot t) \cdot (-1)^k}{\Omega \cdot t - k \cdot \pi}$$

$$= sin(\Omega \cdot t) \cdot \sum_{k \in Z} f\left(\frac{k \cdot \pi}{\Omega}\right) \cdot \frac{(-1)^k}{\Omega \cdot t - k \cdot \pi}$$

That's Shannon's theorem.

We consider $f(t) = sin(a\cdot\pi\cdot t)$ and we set $\Omega = 2 \cdot a \cdot \pi$. If we set $\Omega = a \cdot \pi$, we would get 0. We could choose other $\Omega > a \cdot \pi$, but for $\Omega = 2 \cdot a \cdot \pi$ we see easily that f can be expressed with the Shannon series, even f is not in $\mathcal{L}(R) \cap \mathcal{L}^2(R)$, what is an assumption of the Shannon theorem. With that choice of Ω the coefficients $f\left(\frac{k \cdot \pi}{\Omega}\right) \in \{-1, 0, 1\}$.

$$f_s(t) = sin(2 \cdot a \cdot \pi \cdot t) \cdot \sum_{k \in Z} sin\left(a \cdot \pi \cdot \frac{k \cdot \pi}{2 \cdot a \cdot \pi}\right) \cdot \frac{(-1)^k}{2 \cdot a \cdot \pi \cdot t - k \cdot \pi}$$

$$= \frac{sin(2 \cdot a \cdot \pi \cdot t)}{\pi} \cdot \sum_{k \in Z} sin\left(\frac{k \cdot \pi}{2}\right) \cdot \frac{(-1)^k}{2 \cdot a \cdot t - k}$$

Here is:

$$sin\left(\frac{k \cdot \pi}{2}\right) = \begin{cases} 0 & if \quad k \text{ is even} \\ sign(k) & if \quad |k| = 1,5,9,... \\ -sign(k) & if \quad |k| = 3,7,11,... \end{cases}$$

So we get:

$$f_s(t) = \frac{sin(2 \cdot a \cdot \pi \cdot t)}{\pi} \cdot \sum_{k \in Z} \frac{(-1)^{k+1}}{2 \cdot a \cdot t - (2k+1)}$$

$$= \frac{sin(2 \cdot a \cdot \pi \cdot t)}{\pi} \cdot 2 \cdot \underbrace{\sum_{k \in N_0} \frac{(-1)^{k+1} \cdot (2k+1)}{(2k+1)^2 - (2 \cdot a \cdot t)^2}}_{=\pi/2 \cdot sec(a \cdot \pi \cdot t)}$$

$$= sin(2 \cdot a \cdot \pi \cdot t) \cdot sec(a \cdot \pi \cdot t) \cdot 1/2$$

$$= sin(a \cdot \pi \cdot t) = f(t)$$

References

[1] Ricardo Estrada (1995). *Summability of cardinal series and of localized Fourier series.* Applicable Analysis: An International Journal

[2] J. R. Higgins (1985). *Five short Stories about the Cardinal Series.* American Mathematical Society

[3] Qian, L. (2002). *On the Regularized Whittaker-Koltel'nikov-Shannon Sampling Theorem.* Proceedings of the American Mathematical Society, Vol. 131, No. 4

[4] Schuchmann, M. (2012). *Approximation and Collocation with Wavelets. Approximations and Numerical Solving of ODEs, PDEs and IEs.* Osnabrück: DAV

[5] M. Schuchmann, M. Rasguljajew (2013). *An Approximation on a Compact Interval Calculated with a Wavelet Collocation Method can Lead to Much Better Results than other Methods.* Journal of Approximation Theory and Applied Mathematics (Vol. 1)

[6] M. Schuchmann, M. Rasguljajew (2013). *Approximation of Non $L^2(R)$ Functions on a Compact Interval with a Wavelet Base* (2013, Vol. 2)

[7] M. Schuchmann, M. Rasguljajew; (2013). *Error Estimations in an Approximation on a Compact Interval with a Wavelet Bases.* COMPUSOFT - An international journal of advanced computer technology, Vol. 2, Issue 11, November 2013.

[8] J. M. Whittaker (1927). *On the Cardinal Function of Interpolation Theory.* Proceedings of the Edinburgh Mathematical Society.